UNION INTERNATIONALE DES SCIENCES PRÉHISTORIQUES ET PROTOHISTORIQUES
INTERNATIONAL UNION FOR PREHISTORIC AND PROTOHISTORIC SCIENCES

PROCEEDINGS OF THE XV WORLD CONGRESS (LISBON, 4-9 SEPTEMBER 2006)
ACTES DU XV CONGRÈS MONDIAL (LISBONNE, 4-9 SEPTEMBRE 2006)

Series Editor: Luiz Oosterbeek

VOL. 31

Session WS02

Megalithic Quarrying

Sourcing, extracting and manipulating the stones

Edited by

Chris Scarre

BAR International Series 1923
2009

Published in 2016 by
BAR Publishing, Oxford

BAR International Series 1923

Proceedings of the XV World Congress of the International Union for Prehistoric and Protohistoric Sciences
Actes du XV Congrès Mondial de l'Union Internationale des Sciences Préhistoriques et Protohistoriques

Outgoing President: Vítor Oliveira Jorge
Outgoing Secretary General: Jean Bourgeois
Congress Secretary General: Luiz Oosterbeek (Series Editor)
Incoming President: Pedro Ignacio Shmitz
Incoming Secretary General: Luiz Oosterbeek
Volume Editor: Cris Scarre

Megalithic Quarrying

ISBN 978 1 4073 0405 2

Contacts : Secretary of U.I.S.P.P. – International Union for Prehistoric and Protohistoric Sciences
Instituto Politécnico de Tomar, Av. Dr. Cândido Madureira 13, 2300 TOMAR
Email: uispp@ipt.pt   www.uispp.ipt.pt

BAR Publishing is the trading name of British Archaeological Reports (Oxford) Ltd.
British Archaeological Reports was first incorporated in 1974 to publish the BAR
Series, International and British. In 1992 Hadrian Books Ltd became part of the BAR
group. This volume was originally published by Archaeopress in conjunction with
British Archaeological Reports (Oxford) Ltd / Hadrian Books Ltd, the Series principal
publisher, in 2009. This present volume is published by BAR Publishing, 2016.

Printed in England

BAR
PUBLISHING

BAR titles are available from:

BAR Publishing
122 Banbury Rd, Oxford, OX2 7BP, UK
EMAIL   info@barpublishing.com
PHONE   +44 (0)1865 310431
FAX   +44 (0)1865 316916
www.barpublishing.com

TABLE OF CONTENTS

LIST OF FIGURES

PREFACE

Chris SCARRE

Department of Archaeology, Durham University, South Road, Durham DH1 3LE

The papers in this volume derive from Session WS02 "Megalithic quarries and quarrying: the sources of the stones, their extraction and their manipulation" which was held at the XV Congress of the Union International des Sciences Pré- et Protohistoriques at Lisbon in September 2006. Megalithic monuments are a key and prominent feature of the Neolithic archaeology of western and northern Europe, where their origins can be traced to the 5th and 4th millennia BC. They include famous sites such as Stonehenge, Carnac and Newgrange, alongside many thousands of smaller chambered tombs, stone rows and stone circles, down to the isolated menhir which may measure only a metre or so in height. Their numbers are often impressive. A late 19th century inventory in Denmark recorded 7.287 megalithic tombs (of which 2.364 were visible upstanding monuments), and more recent estimates suggest these may be only c.10% of the original, implying that some 23.000 megalithic tombs once existed there (Tilley 1996, 130). Hence in some regions of Europe, megalithic monuments may have been a remarkable and ubiquitous feature of Middle and Later Neolithic landscapes.

Megalithic monuments are distinguished from other types of prehistoric structure by their incorporation of large (or as Childe put it "extravagantly large") slabs of stone. The tradition of megalithic architecture was far from opportunistic but is highly indicative of the attitudes of these societies to the materials that they were using. As the following studies show, the stones were not quarried from a depth but came from surface exposures or boulder scatters. Hence the sources were visible in the landscape before construction was even contemplated, and the megalithic blocks linked these monuments directly to those landscape features. Yet the megalithic builders drew on the landscape not only for the materials that they required (the large stone slabs and other elements), but also for the memories, powers and associations that those materials incorporated and symbolised.

The decision to take 'megalithic quarries' as the subject of this UISPP session was not driven by a concern with the technical aspects of the quarrying process, but by the implications of stone sources for broad issues of social organisation, cosmology and landscape. By comparison with other aspects of megalithic monuments, the manner in which blocks were taken from rock outcrops, boulder spreads, or other sources has been a somewhat neglected domain. That contrasts with the relatively routine discussion of associated issues such as megalithic transport, or the relationship of sites to landscape features. The idea that megalithic blocks were taken from places already endowed with sacred or mythological significance is a prominent element in many recent archaeological studies of megalithic monuments, particularly those undertaken by British prehistorians, yet is only rarely accompanied by specific attention to the source locations themselves. Recent years have, however, seen the identification of a number of megalithic quarry sites, both in Britain and overseas, and in a few cases these have been followed up by detailed analysis and excavation. The results allow us to be more precise about the land forms from which the blocks were taken, and the manner in which they were extracted; to determine whether, for example, natural weathering processes had already pre-formed the blocks that were removed.

The papers in this volume thus complement recent studies of the materiality of megalithic monuments which have focused on issues of texture or colour (e.g. Cummings 2002; Jones 1999; Trevarthen 2000; Scarre 2004). They demonstrate the insights to be gained by identifying how and from where megalithic blocks were taken. They can reveal the nature and appearance of 'pre-megalithic' landscapes, and the character of the stone sources which in some cases were destroyed by the very process of building the megalithic monument. Such studies allow a new appraisal of the beliefs relating to stone that lay behind the concept of megalithic construction.

Acknowledgements

I would like first of all to thank Professor Luiz Oosterbeek and his team for their organisation of the UISPP meeting, for inviting me to participate, and for accepting this session in the programme. I am also grateful to the speakers who have contributed to this volume. Special thanks are offered to Emmanuel Mens for his assistance with several of the French abstracts.

Bibliography

CUMMINGS, V. (2002) Experiencing texture and transformation in the British Neolithic. *Oxford Journal of Archaeology* 21, 249-261.

JONES, A. (1999) Local colour: megalithic architecture and colour symbolism in Neolithic Arran. *Oxford Journal of Archaeology* 18, 339-350.

SCARRE, C. (2004) Choosing stones, remembering places: geology and intention in the megalithic monuments of western Europe. In Boivin, N.; Owoc, M.A., eds. *Soils, Stones and Symbols. Cultural perceptions of the mineral world*. London: UCL Press, 187-202.

TILLEY, C. (1996) *An Ethnography of the Neolithic. Early Prehistoric Societies in Southern Scandinavia*. Cambridge: Cambridge University Press.

TREVARTHEN, D. (2000) Illuminating the monuments: observation and speculation on the structure and function of the cairns at Balnuaran of Clava. *Cambridge Archaeological Journal* 10, 295-315.

Caption to cover illustration

Rock outcrops and tumbled blocks at Carn Menyn in the Preseli Hills of southwest Wales, source of some of the Stonehenge 'bluestones'

STONY GROUND: OUTCROPS, ROCKS AND QUARRIES IN THE CREATION OF MEGALITHIC MONUMENTS

Chris SCARRE

Department of Archaeology, Durham University, South Road, Durham DH1 3LE

"That they preferred to use large ones for certain purposes was not due to ignorance or chance, it was because the large stone as such had some particular meaning and association for them. We cannot definitely say that large stones were themselves actually worshipped, but there can be no possible doubt that for some reason or other they were regarded as peculiarly fit to be used in sanctified places such as the tombs of the dead."

T. Eric Peet *Rough Stone Monuments and their Builders* 1912, 150

Abstract: The inherently peculiar nature of megalithic architecture arises from the employment of large stone slabs that were frequently unmodified and unshaped. Subsequent studies of the megalithic slabs themselves have focused mainly on their geological origin and the distances over which they were transported. The way that the slabs were extracted from their source material has been only rarely addressed, although megalithic 'quarries' have occasionally been identified. The deployment of glacial boulders in North European monuments is a well-known phenomenon, and even beyond the glaciated zone, extensive spreads of natural boulders may have characterised large areas of western Europe during the earlier Neolithic. Some megalithic monuments were built directly from such scattered blocks. In the majority of cases, however, the megalithic slabs that were used can be shown to have been cut away from cliffs and outcrops, exploiting natural fracture planes. Whether quarried slabs or detached boulders, what unites these sources of stone is that they were surface exposures, visible features of the early Neolithic landscape that may already have been places of special significance.
Key-words: megalithic monuments, megalithic transport, quarrying, sarsens

Résumé: Le caractère intrinsèquement singulier de l'architecture mégalithique se distingue par l'emploi de gros blocs de pierres restés très souvent "brut de débitage", sans modifications, ni façonnage. Les études plus récentes sur les dalles mégalithiques se sont concentrées plus particulièrement sur l'origine géologique des pierres et sur leur distance de transport. Le processus d'extraction de ces dalles a beaucoup moins retenu l'attention des scientifiques, malgré la découverte ponctuelle de 'carrières' mégalithiques, et ce fait bien connu qu'est l'emploi des rochers d'origine erratique dans les monuments d'Europe du nord. Même au-delà de la zone touchée par les glaciers quaternaires, il est probable que des surfaces parsemées de blocs naturels détachés ont dû caractériser les grandes étendues des paysages ouest-européens au début du Néolithique. Quelques monuments mégalithiques ont été construits directement avec ces blocs détachés. Pour la plupart cependant, les dalles mégalithiques employées dans ces monuments semblent avoir été extraites de falaises ou de rochers en exploitant les failles naturelles. Tant pour les dalles extraites de cette façon, que pour les blocs détachés, ce qui semble caractériser toutes ces pierres, c'est le fait qu'elles proviennent de gisements superficiels, qui pour la plupart ont dû constituer des éléments particulièrement visibles dans les paysages néolithiques et que les sociétés préhistoriques avaient peut-être déjà investi d'une signification particulière.
Mots-clés: monuments mégalithiques, transport des mégalithes, carrières, sarsens

The word 'megalithic', it is well known, derives from the Greek *megas* large and *lithos* stone. It appears first to have been used in 1849 by Algernon Herbert, and the related term 'megalith' for the individual block of stone, was added in 1853 by Frederick Collins Lukis in the title of his paper 'Observations on the Celtic megaliths' (Herbert 1849; Lukis 1853).[1] By the time these words came into use, megalithic monuments had already attractted a considerable scholarly literature. The 'megalithic' terminology drew particular attention to the size of the stones, and coincided with the first references to the processes of extraction and construction that must have been employed. The Baron de Bonstetten, in his famous 'Essai sur les dolmens' noted how in northern Europe, the megalith-builders made use of glacial erratics. He notes

that in other regions, too, they could have used detached blocks of stone, but that since the latter would not have been sufficiently abundant, the builders must commonly have had to extract the blocks from bedrock using hammerstones, wooden wedges and fire-setting (Bonstetten 1865, 23). The concept of megalithic quarrying was hence introduced, but was developed little further in the decades that followed. It is in effect only during the last thirty years, and mainly within the last ten, that attention has turned to identifying and studying the precise sources and outcrops from which the blocks were derived.

While megalithic quarries are a relatively recent focus of study, the origins and petrology of megalithic slabs have been analysed and discussed almost from the beginning of scholarly interest in the monuments themselves. The Astronomer Royal, Edmund Halley, noted in 1720 that the stones of Stonehenge were of two distinct lithologies, and in 1923 these were finally traced to their sources on

---

[1] The latest edition (including the online edition) of the *Oxford English Dictionary* correctly attributes to Herbert the invention of 'megalithic' but mistakenly dates his publication *Cyclops Christianus, or an Argument to disprove the supposed Antiquity of Stonehenge and other Megalithic Erections in England and Britanny* to 1839.

the Marlborough Downs and the Preseli Mountains respectively (Chippindale 1994, 79; Thomas 1923; Darvill this volume). Most megalithic monuments, however, derive their materials from within a much narrower compass. Frederic VII, king of Denmark, indeed, went so far as to argue that megalithic monuments must have been built in those very places where the megalithic blocks themselves lay lying on the surface, and that the chambers were simply spaces dug out beneath such natural blocks (1857, 4). Though he subsequently abandoned the idea, this argument is curiously reminiscent of recent discussions about the power of ancestral places.

In addition to the sources of the blocks, and the associated question of what we may call 'megalithic transport' (the means by which the stones were dragged to the sites), there is the issue of tooling and shaping. In many megalithic monuments the method of construction is characterised not only by the size of the stones but also by the fact that they were left largely unshaped. In many cases it is evident that little attempt was made to trim the blocks into shape, or to smooth and straighten their edges and surfaces. The 'brute stone' nature of the tradition was highlighted by 19th and early 20th century writers. Peet, for example, defined a megalithic monument as "usually, though not quite invariably, made of large blocks of unworked or slightly worked stone" (Peet 1912, 2). The observation had implications for the manner in which the materials had been obtained, drawing attention to "the great effort that they [the builders of these monuments] were prepared to make to avoid quarrying" (Kendrick 1925, 102). Recent discussions of the materiality of megalithic blocks have revived interest in this issue. The use of large, generally unshaped stones is indeed no 'primitive' architecture but the consequence of specific choice and tradition. Above all, the reluctance to fracture and shape the stones, and the key importance of specific types of sources including cliffs, outcrops and boulder fields, indicate a significant symbolic dimension to the selection and use of these materials that gives the concept of 'megalithic' architecture an explicit cultural meaning.

The social and symbolic significance of megalithic blocks is revealed by ethnographic evidence from those few regions of the world where traditional practices of megalithic construction have continued into recent times. One such is island southeast Asia. Adams (this volume) draws attention to the contrast between the megalithic tombs of Sumba, where the stones are carefully shaped and carved, and those of the Toraja on Sulawesi, where the megalithic blocks are not quarried but taken unmodified from hills and streams. In the latter case, stones of the desired shape are obtained not by shaping but by systematic searching among the naturally available forms. Elsewhere in southeast Asia, the significance of stone is traditionally interpreted within an animistic context. The Maloh of Sarawak, for example, consider stone a potential receptacle for spirits and pay particular attention to stones that have the recognisable form of a human being or animal. Smaller stones are collected and

curated and used in curing rituals; larger masses of stone are associated with mythological explanations and may become objects of worship. Isolated rocks and outcrops are considered especially significant (King 1976). As King emphasises, "The occurrence of stone in the landscape ... is not explained in geomorphological or geological terms, but in supernatural and mythical terms" (King 1976, 107).

There is little archaeological evidence for the rituals that may have accompanied the quarrying of megalithic blocks, but other ethnographic parallels indicate that the extraction of material from the earth is commonly associated with special practices to propitiate the spirit-owners. That may be especially the case where animistic readings prevail. The relevant ethnographic accounts concern quarrying not for megalithic blocks but for stone axes, clay or metals. They nonetheless provide insights into traditional understandings of natural materials. In Australia and Papua New Guinea axe quarries were considered to be sacred locations that could only be approached once certain rituals had been fulfilled (Boivin 2004, 11). Ethnography indicates that mined rock outcrops in Australia were almost invariably considered the creation of powerful Ancestral Beings (Brumm 2004, 147). The Tungei of Papua New Guinea believed that the axe stone was controlled by two spirit sisters, to whom pigs and a mythically important giant rat were sacrificed at the beginning and end of a quarrying expedition (Burton 1984, 239-240). In North America, the annual working of the catlinite quarry at Pipestone, Minnesota by groups of Sioux was preceded by three days of fasting, prayers and sacrifices (Scott & Thiessen 2005).

Such practices leave few archaeological traces but there is little doubt that prehistoric quarrying was similarly entangled with belief systems telling of supernatural beings and appropriate ritual practices. In northwest England, the axe factories of Great Langdale exploited seams of rock that were marked out by their inaccessibility. It may have been the sacredness of particular rock sources that gave the axes produced from them especial potency, and encouraged this venture into a remote and difficult location (Bradley & Edmonds 1993). The quantities of antler picks left on the floors of Neolithic flint mines in southern and eastern England provide archaeological evidence that the winning of materials from the ground was accompanied by rituals of deposition. In this case, the practice appears to have marked the abandonment of the shafts, as the placement of the antler picks inhibited any further working (Barber et al. 1999, 66). At Lambay off the east coast of Ireland, excavation of the Eagle's Nest axe factory has revealed how quarry debris was returned to the earth by deposition in pits, returning unused material to the land, perhaps to replenish the ancestral power of the place and ensure its continuity (Cooney 2000; 2005).

Ethnographic and archaeological studies of the use of stone by non-Western societies promote the concept of material agency as a means of better understanding the use of megalithic blocks (Gell 1998; Scarre 2009). The targeted selection of blocks drawn from particular places in the landscape, and their careful and intentional arrangement within a stone setting or chambered tomb, suggests that individual megalithic slabs may have been ascribed with their own identities by the megalith-builders. In some cases, this is compounded by the anthropomorphic shape of the stones or (more rarely) by the addition of human features such as pairs of breasts. Even without such direct visual cues, however, tall, narrow standing stones are easily envisaged as fossilised humans, an identification which finds ample illustration in the popular names of monuments such as 'the Merry Maidens' or 'Long Meg and her Daughters'. Indeed, the attribution of human qualities to animate and inanimate objects is a pervasive feature of all religion (Guthrie 1993; Boyer 1996).

Megalithic slabs were special, then, not only on account of their size, but also for their associations and connotations. They formed part of a symbolic understanding of the natural world by prehistoric communities in which stone held a key significance, one that was emphasised wherever that normally hidden underlayer of the natural world was exposed in the form of cliffs, outcrops or boulder spreads. In this light, the construction of megalithic monuments takes on a new meaning, one that is explored in different ways in the papers in this volume. Questions of megalithic transport and megalithic construction will be touched upon, but the focus here is on the quarrying and extraction of the blocks. A diversity of practices is revealed, with some stones obtained from surface scatters of detached blocks; others by the removal of blocks from exposed cliffs or outcrops; or more rarely by quarrying from a greater depth. We may assume that all such activities in a prehistoric context were enmeshed within relations of meaning that required and inspired ritual action. The absence of material evidence for such ritualisation in the context of megalithic block extraction may simply reflect the fact that very few such quarries so far have been identified, and still fewer of them excavated.

TRACING THE SOURCES

It is stating the obvious to observe that the availability of suitably sized stones is a key requirement of megalithic construction, and impacts heavily on the distribution of megalithic monuments through western and northern Europe. The sources that were drawn upon were, however, diverse. First there are the petrologies themselves, with granites and schists common in western Iberia, northwest France and southwest Britain, and in the North European Plain, whereas limestones and sandstones were used in Orkney, central southern Britain, the Paris basin, southern France and southern Iberia. These materials have variable working properties and constrain in some respects the size and shape of the blocks: the

characteristic tall thin pillars with sharply oblique tops of the Stenness and Brodgar circles derive directly from the fracture patterns within the beds of Orkney flagstone.

Such patterns and properties at once create a visual link between megalithic monuments and stone sources, and the association is reinforced by qualities of texture and colour. Striking contrasts are sometimes achieved that demonstrate that the builders of these monuments were not only sensitive to these differences but employed them in the creation of positive material statements. At Saint-Just in Brittany, the three rows of standing stones that constitute the Le Moulin alignments comprise two lines formed exclusively of quartzite blocks (file ouest and file nord), and a third combining the brilliantly white quartzite with pillars of brown or dark grey schist (Le Roux et al. 1989; Scarre 2002). The fact that the quartzite was brought to the site from a distance of 4 kilometres underlines the intentionality behind this arrangement. At Newgrange in Ireland the reconstructed façade of quartz and granitic cobbles stands above a ring of greywacke kerbstones. Analysis of the quartz shows that most if not all came from the Wicklow Mountains some 60 kms south of Newgrange, while the granitic material has been sourced to the Mourne Mountains 35 kms to the north. Hence the use of these two materials could be interpreted as symbolizing the Neolithic landscapes to the north and south of the passage tomb and their convergence at it (Meighan et al. 2002, 2003). The current reconstruction has been much criticised, and it is unlikely that the Newgrange façade ever took the form of a near-vertical wall of white peppered with black (Eriksen 2004, 2006), yet even if the stones were originally spread across the surface of the mound, or formed a series of platforms at its foot, the striking colour contrast of the two materials would have been highly conspicuous.

In other examples, the visual contrast between different materials would have been less obvious, especially where they were built into enclosed burial chambers which can only have been viewed in conditions of low light. It is clear nonetheless that the deployment of materials of different origin was in many cases structured and intentional. At Carn Ban on Arran, red schist from the south of the island was used for the dry-stone work of the chamber and for two of the four capstones (in alternation), while the façade, the half-height orthostats within the chamber and the other two capstones are of white schist from the north (Jones 1999). Similar colour patterning has been observed at the Clava cairns in northeast Scotland (Trevarthen 2000). Nor are such phenomena restricted to Britain and Ireland. At Pornic in northwest France, patterned colour contrasts between the megalithic slabs used in the construction of chambered tombs were noted by 19th century antiquarians, notably in the Caveau de la Croix and Les Trois Squelettes (De Wismes 1876, 208; Lisle du Dréneuc 1882, 268; Scarre 2004).

The sources of the megalithic blocks may readily have been recognized by prehistoric individuals familiar with

Fig. 1.1. Anta da Lajinha (Portugal): schist orthostats of megalithic chambered tomb (above) and columnar schist outcrops (below) from which these slabs may have been extracted

the cliffs, outcrops and boulders of their local terrain. In inland Portugal, the Anta da Lajinha consisted originally of a chamber of eight schist slabs leaning inwards and resting against each other in the manner typical of this region. While only two of the slabs remain, there are sufficient fragments of the others to indicate that the material was in each case the same. The contrast here is not between the petrologies of different orthostats but between the orthostats and the bedrock; for whereas both are of schist, the bedrock below the tomb is highly fissured and incapable of yielding blocks of any size. The source of the tabular plate-like orthostats can however immediately be recognised in the rows of pillar-like outcrops that traverse the adjacent hillsides. In colour, size and texture these bear a striking resemblance to the schist orthostats of Lajinha, and their vertically fissured form makes it easy to visualise the manner and ease with which individual slabs could have been detached, albeit no direct trace of working has been observed on the outcrops closest to the site (200 metres to the east) (Fig. 1.1). The monumental nature of the outcrops themselves suggests a symbolic dimension in the relationship of structure to stone source, and would have been visually reinforced if (as believed from recent excavation) the mound at Lajinha did not entirely cover the chamber but would have left the orthostats and now-missing capstone openly visible.[2]

In addition to colour contrasts and visual qualities, the selection and deployment of materials of different character and origin may sometimes have been driven by their varying mechanical properties. At the Bougon cemetery in western France, for example, a complex of five cairns contains a total of eight burial chambers of mixed megalithic and dry-stone construction, all of limestone. The dry-stone material was quarried on site, but the larger slabs are of three different varieties of limestone: Lower Callovian, Corallian, and Bathonian with flints, all available within a 4 km range. The differ-rent types of limestone were drawn upon preferentially for different structural elements. The majority of the ortho-stats, for example, are of fine-grained Lower Callovian whereas the much coarser Corallian and Bathonian furnished further orthostats but were used more particularly for the larger capstones (Cariou in Mohen & Scarre 2002). This deployment of materials can be viewed in terms of their mechaniccal properties. The Corallian and Bathonian are both tough, dense materials highly resistant to erosion and with good load-bearing capacities. These characteristics made them ideal materials for the 32-tonne capstone of chamber F2 and the 90-tonne capstone of chamber A. They were worked into the desired shape only roughly, if at all. Many of the orthostats, by contrast, were carefully smoothed by pecking. This is particularly noticeable in chambers C and F2, where 'crochets' or 'hooks' in raised relief were left projecting from the surfaces of some of the slabs. The shaping of the four side slabs of chamber C (originally

probably a closed box-like cist) was particularly extensive: the base of the eastern wall slab was carved into a curved projecting lip that fitted against the edge of the floor slab of the chamber; while the northern wall slab had grooves worked in both edges that engaged precisely with the ends of the adjacent eastern and western wall slabs.

Thus both colour contrasts and structural considerations may have been among the reasons for the use of material from different sources in megalithic monuments. Another was the availability of suitably sized blocks. The three surviving tombs of the Vale de Rodrigo in the Alentejo region of Portugal, for example, were built primarily of biotite-tonalite slabs, a material which occurs some 3 kms to the west and north of the monuments and which has fracture planes that readily yield usable megalithic blocks. The tombs also contain orthostats of biotite-tonalite-hornblendite, a material very similar to biotite-tonalite and one which occurs closer to the tombs to the east, but owing to natural fracture patterns does not yield many usable blocks. The preference for one over the other is thus easily explained by their suitability as sources of megalithic blocks. It is more difficult to explain why all three tombs also incorporate porphyritic granodiorite, which was brought from much greater distances, probably of at least 7 kms, especially as this seems to be a more brittle material (Dehn *et al*. 1991; Larsson 1998).

Kalb remarks that the presence of blocks from distant sources is characteristic of the large monuments of Vale de Rodrigo, while the smaller monuments usually use local material and often of only one type (Kalb 1996). A similar correlation between tomb size and petrological complexity has been noted in the Channel Islands. In a series of studies, Mourant documented the variety and sources of the materials used to build the megalithic monuments of Jersey (Mourant 1933, 1939, 1963, 1977). Significantly it was the largest of these, the passage grave of La Hougue Bie, that incorporated both the largest number of stone types, and those that had been transported over the greatest distance. Thus whereas Faldouet, Le Mont Ubé, Le Mont de la Ville and La Sergenté were built of materials from within a distance of 2 kms, La Hougue Bie drew on materials from seven sources, including one slab from Mont Mado on the opposite side of the island (Mourant 1933) (Fig. 1.2). This pattern has been interpreted in social terms, as the reflection of a hierarchy in which Faldouet and the other smaller tombs may have been built by local communities, whereas the building of La Hougue Bie drew together communities from across the whole of eastern Jersey (Patton 1992). More recent work has extended the discussion to include the Guernsey monuments, based on a new set of petrological identifications (De Pomerai 1997). It is clear that the patterning of stone in Channel Islands monuments cannot be explained by either the local availability of materials, nor by their engineering properties, but was driven by social or symbolic considerations. The stones that were transported over considerable distances may have been tokens of local

---

[2] Excavations at the Anta da Lajinha 2006-8 were conducted by the author together with Professor Luiz Oosterbeek of the Instituto Politécnico de Tomar.

La Hougue Bie (Jersey): petrology of the megalithic blocks

syenitic granite
Fort Regent red granite
La Rocque coarse granite
Mont Mado granite
Le Hocq fine granite
coastal diorite
granite porphyry

0   2m

Jersey: sources of megalithic blocks

1 La Hougue Bie
2 Faldouet
3 Le Mont Ubé
4 Le Mont de la Ville
5 Les Cinq Pierres
6 La Sergenté
7 Les Montes Grantez
8 La Hougue des Géonnais

0   5 km

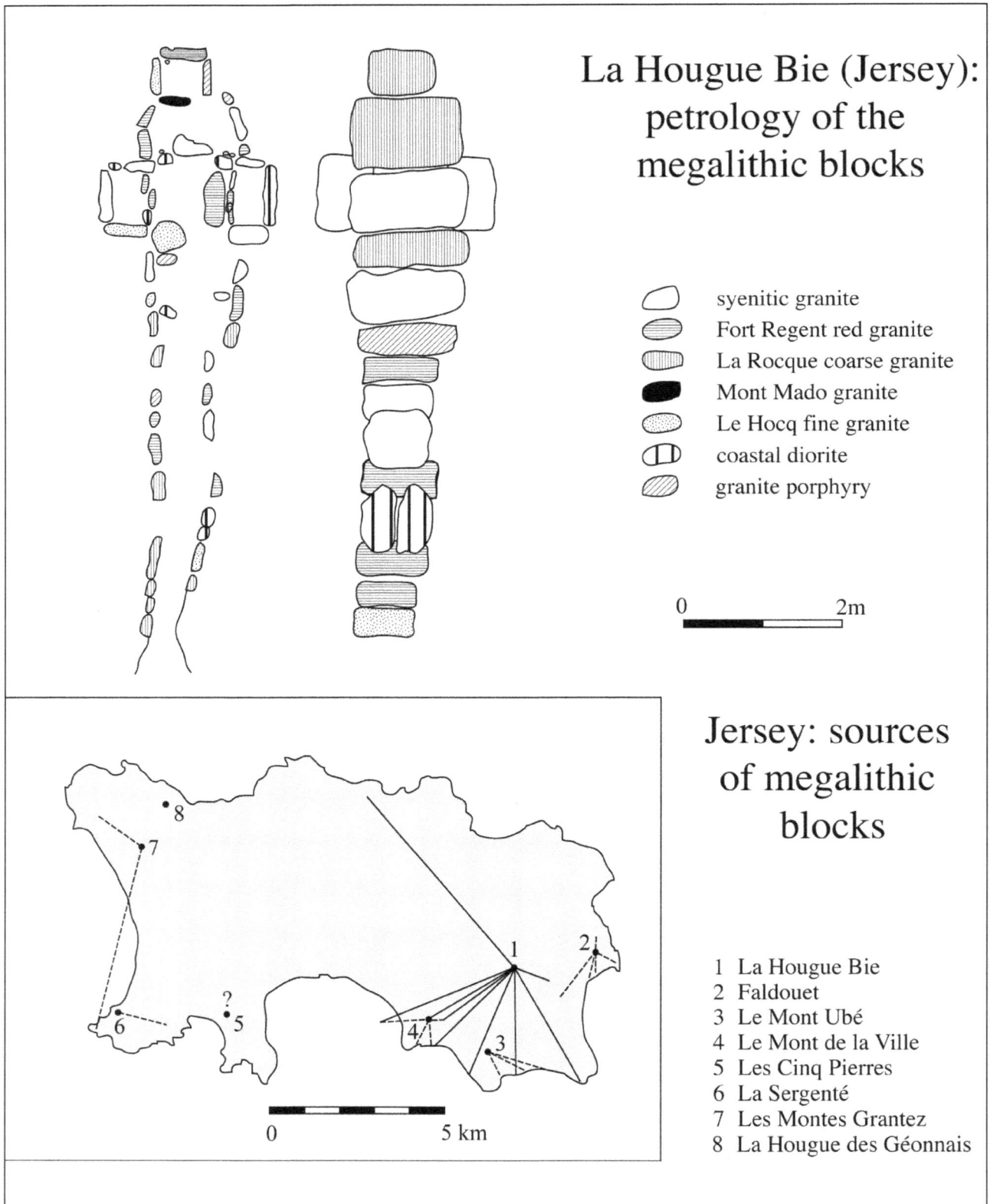

Fig. 1.2. Megalithic transport on Jersey:
(above) differing petrologies of megalithic blocks at La Hougue Bie (after Mourant 1933, Patton *et al.* 1999);
(below) stone sources and transport distances (after Patton 1992)

identity, and the placement of stone types in particular positions within a tomb could have been the means used by passage grave builders to negotiate power and compete for prestige (Bukach 2003, 31).

Alternatively, we might consider the identities of the megalithic blocks themselves to have been the active principle. That could explain the distances from which they were transported. What was so very special about the stones, or the cliffs and outcrops from which they were taken, to lead communities to transport them over such large distances? The incorporation of one or more blocks from distant sources, picked out from the rest by their distinctive colours and textures, may be compared to the re-use of former menhirs in passage graves, a practice that has been extensively documented in north-west France but has also been recognised in Iberia, Britain and Ireland (L'Helgouach 1983, 1997; Bueno Ramírez & Balbín Behrmann 2007; O'Sullivan 2009; Richards 1996). The common element in both cases may be the targeted use of monolithic blocks that had already acquired special sanctity and significance, whether as part of an outcrop or boulder field, or part of an earlier megalithic monument.

We should at this point return to the issue of quarrying. Long-distance megalithic transport is a relatively unusual phenomenon and most megalithic monuments were built of materials from the local area, both in southern and northern Europe (Kalb 1996; Bakker & Groenman-van Waateringe 1988; Thorpe & Williams-Thorpe 1991). What long-distance transport implies is that in some cases at least the specific stone sources that were selected had an importance in their own right, and were not just the closest or most convenient places from which suitable blocks could be derived. Thus the study of megalithic quarrying must not be reduced to a functionalist account of least-effort principles, and analogies with the mining of flint or the circulation of polished stone axes highlight the social and symbolic dimensions of raw material exploitation by Neolithic communities. The focus of this volume, however, is not only the symbolism of megalithic blocks, but the archaeology of their extraction.

SURFACE FEATURES AND VANISHED WORLDS

The building of megalithic monuments in western Europe took place in a landscape very different from that which we see today. Many areas would have had extensive forest cover. This was interrupted by clearings, coastlines and river valleys and was perhaps not as uniform as has sometimes been believed (French *et al.* 2003; Allen 2005), but pollen profiles in general show that clearance by early farming communities was a gradual process extending over many centuries. In Brittany, forest clearance gathered pace only during the 3rd and 2nd millennium BC, especially in inland areas, and even in the heavily monumentalised southern Morbihan charcoal from buried soils suggests that dense stands of oak survived in certain places (Marguerie 1992; 2000). Hence many megalithic monuments may have been constructed in landscapes that had not wholly been cleared for agriculture, although the evidence of Céide in northwest County Mayo, where tombs were constructed within a mid-4th millennium field system (Caulfield *et al.* 1998), warns us against the danger of over-generalisation.

Landscapes as yet uncleared for agriculture will have been littered with rocks and boulders that have long since been removed, not least for use in stone walls and buildings. Many megalithic monuments have suffered in the process. A classic case is the Baltic island of Rügen, where the 229 megalithic monuments recorded in 1829 had been reduced to only 38 by 1929 (Sprockhoff 1938, 50-51). The systematic clearance of rocks, boulders and many monuments from the landscapes of western and northern Europe largely obscures the relationship of megaliths to an important category of source material: detached blocks and erratics. In the North European Plain, glacial erratics deposited in the end moraines left by retreating ice sheets constituted the primary source of megalithic blocks. In Mecklenburg and Pomerania, there is a clear spatial association between megalithic tombs and successive end moraines (Gehl in Schuldt 1972; see especially distribution maps 13 & 15).

A similar connection can be made at the western end of the North European Plain. Here the Drenthe *hunebedden* lie within the area of a Saalian end moraine, but Bakker has stressed the distorting effect of historical stone clearance on the distribution of suitable erratic material, and on our impressions of the landscape in general (Bakker this volume; Bakker 1992, 36; Bakker & Groenman-van Waateringe 1988). A particularly intense phase of clearance occurred in the 18th century, when pile worm attack forced Dutch engineers to rebuild timber revetments along canals and quaysides in stone, placing increased pressure on the limited sources available. Nonetheless, some very large blocks, such as the Tryglaw erratic (or 'Triglaffstein') in Pomerania, weighing almost 2000 tonnes, have survived. Within Drenthe, piles of smaller stones at Lage Vuursche and Hengelo have mistakenly been identified as hunebeds, while Van Giffen identified the large block at Noordbarge as an erratic boulder of Norwegian granite (Van Giffen 1927, 139)

The boulder fields of the North European plain would have been dramatic and extensive features once erosion processes freed the many erratics from the boulder clay in which they were originally encased. Yet landscapes littered with scattered blocks of stone were far from being restricted to the glaciated regions of Europe. The sarsens that provided the primary constructional material for many megalithic monuments in Wiltshire and adjacent counties were drawn from scatters of blocks that formerly covered the chalklands of southern England from Kent to Dorset (Fig. 1.3). They were formed from Tertiary sands deposited over the chalk downlands, and were locally hardened by silcrete formation to create sarsen boulders (Geddes & Walkington 2005, 62-63). The Marlborough Downs are commonly cited as the likely source for the Avebury megaliths and the larger stones at Stonehenge, but it is clear that sarsen blocks were once considerably more abundant and widespread across this landscape,

Fig. 1.3. Natural scatter of sarsen blocks at Lockeridge Down near Avebury (Wiltshire). (Photo: Timothy Darvill)

making the precise sources impossible to determine. Some of the stones may have lain close to hand, or even have been taken from the sites themselves (Field 2005, 91). Scattered sarsen blocks, traditionally called 'greywethers' from their resemblance to sheep, can measure up to 4 metres in length (Darvill 2006, 22). Several tens of thousands of these blocks are estimated to survive (Geddes & Walkington 2005, 63), though over the course of time, the larger blocks that were formerly available have been exhausted through the combined effects of agricultural clearance and building activity. None of the surviving surface stones approaches the dimensions of the Stonehenge sarsens (the largest, Stone 56, measuring c.9.7m: Walker in Cleal *et al.* 1995, 29), but a 7-metre long sarsen was discovered from a sinkhole at Aston Rowant in Oxfordshire (Bowen & Smith 1977, 189). Some of the dressed sarsens of Stonehenge may have come from such sinkholes, since buried sarsen is relatively soft but hardens on weathering; if carved when fresh it would have been easier to work (Geddes & Walkington 2005, 63).

To prehistoric communities, the sarsen blocks scattered across the Wessex landscape may have attracted supernatural explanations and have become objects of awe and wonder, "rocks with their own ancient and living history" (Field 2005, 89). Such a dense litter of symbolically charged boulders may well have been the inspiration for the first megalithic monuments. As Field has remarked, the process could have begun through the imitation of groupings of stones that were present as natural phenomena (*ibid.*, 89). The sites of Stonehenge and Avebury themselves may have had to be cleared of

boulders before the monuments were laid out, and it is possible that some of the cleared stones were incorporated in the megalithic structures that were built on those sites. Unmodified sarsen boulders may have possessed their own special significance. Nineteenth-century field clearance at Winterbourne Monckton and Overton Down revealed that several of the larger sarsens covered or marked pits containing multiple inhumations, in some cases associated with Beaker material (*ibid.*, 92).

In northwest France, too, extensive scatters of surface stones once existed, and documentary evidence suggests that much of the clearance may be of recent date. Local surveys undertaken in 1733 indicate that no less than 46% of the land area of Brittany was uncultivated (Sutton 1977), and scatters of blocks were a common feature of the landscapes explored by early antiquarians. The English cleric John Bathurst Deane, who explored the Carnac alignments in 1832, noted in one place "a very remarkable bed of rocks, which had the appearance of having been a quarry for the Dracontium. Some large stones were lying loose upon its surface, as if they had been prepared but never erected" (Deane 1834, 209).[3] This observation led him to suggest (probably erroneously) that the monument had never been completed as intended. A local hotelier informed him that within his recollection, some 1500 to 2000 stones had been removed between Carnac and Sainte-Barbe (Deane 1834, 211). The quantity of scattered surface blocks in

[3] Deane, following Stukeley's interpretation of Avebury (Piggott 1985, 104-5) was inspired by the idea that the Carnac alignments had originally been constructed in the shape of a serpent, and were a *Dracontium* (snake or dragon temple).

this region must originally have been considerable. Some thirty years later, Geoffroy d'Ault-Dumesnil argued that the stones used to build the Carnac monuments were found on the surface, not taken from quarries, and still less brought from a distance. He observed among the surviving scatters of granite boulders several natural formations that resembled dolmens and menhirs, concluding that the builders had used only the materials that Nature had prepared for them (d'Ault-Dumesnil 1866).

Deane noted that, whether loose rocks or outcrops, the sources of the Carnac stones lay close to the alignments themselves. This has been confirmed by recent studies. Dominique Sellier has demonstrated how the weathering patterns and distribution of stone sizes in the Kerlescan stone rows correlates spatially with the fracture patterns in the bedrock. Some of the standing stones may have been detached blocks; the remainder could have been separated from the bedrock with relatively little effort (Sellier 1991, 1995) (Fig. 1.4). For the Kermario alignments, on the other hand, the blocks were quarried from an outcrop or outcrops that were entirely removed in the process (Mens, this volume). Indeed, *pace* d'Ault-Dumesnil, it has now been demonstrated that the majority of blocks used in the megalithic monuments of the Carnac region were not detached boulders but had been quarried from outcrops (Mens, this volume). But adventitious exploitation of surface material was certainly practised, as Sellier has demonstrated: both methods were used. Mens' study has shown how it is possible to distinguish within a single monument those blocks that came from superficial deposits (that display weathered surfaces) and those from within outcrops (without traces of weathering), even where the oucrops from which they were derived have disappeared. His work suggests that quarrying may have been resorted to only when all the suitable loose blocks or easily detached blocks had been exhausted (Mens 2007). It is likely that other factors also were involved, however, since for prehistoric societies the natural features of the Carnac area probably carried a diversity of meanings and associations, with individual boulders and outcrops perhaps invested with special sacred significance that made them particularly appropriate for specific megalithic monuments.

There are other cases where the construction of a Neolithic monument may have been directly connected to the presence of surface stones. Evidence suggests that earth-fast boulders may sometimes have been dug out and raised on pillars to become the capstones of megalithic chambers built within or around the very hollow from which the stone had been removed. The pit below the small polygonal chamber of Carreg Samson in southwest Wales was noted by the excavator as the place from which the capstone may have been dug (Lynch 1975, 16; see also Richards 2004a). A similar pit beneath the Pentre Ifan portal tomb could be interpreted in a similar way. Whittle has added to this suggestion the observation that these are structures with massive capstones raised on

conspicuously slender pillars, conjuring the evocative image of 'stones that float to the sky' and leading us to wonder whether the purpose of these structures was not to create a closed funerary chamber but to venerate and display the capstones themselves (Whittle 2004). This recalls the early suggestion that the builders of megalithic tombs, unable to move blocks of megalithic dimensions, were obliged to "select the location where the large stone that was destined to be the capstone was already in place" (Frederic VII 1857, 4).[4]

Some megalithic monuments went further than this and exploited or incorporated *in situ* natural boulders. The Wiltshire long mound of South Street was built over a cluster of nine sarsen boulders, which may have been a pre-barrow shrine, occupying the position beneath the barrow where a mortuary house would normally be found (Pollard & Reynolds 2002, 61; cf. Whittle 1993). Sarsen boulders were also incorporated in the base of Silbury Hill though whether these were natural blocks *in situ* or had been intentionally placed in position remains unclear (Field 2005, 91). In southwest Wales, burial chambers were sometimes created simply by levering up a natural slab to form a space beneath. Sites such as Carn Gilfach and Carn Wnda are difficult to distinguish from the boulders and outcrops that litter the surrounding hillsides (Barker 1992; Cummings 2002) (Fig. 1.5). Megalithic monuments in western France occasionally incorporate natural outcrops in their construction. At Brécé in Mayenne, for example, the southeastern end of the chamber was partly quarried into a rock outcrop which formed that end of the chamber (Bouillon 1989). The megalithic tomb on the small island of Enez Bihan was also built on a rock outcrop and a natural boulder forms most of the northwest side of the chamber (Daire & Le Page 1994). Similarly at Kerherne-Bodunan and Pont-Bertho in the Landes de Lanvaux, a natural outcrop forms one side of the elongated chamber (Gouézin 1994, 86-87; 107-108). This incorporation of natural outcrops in the structure of the tomb finds its most extreme expression at Men-Guen-Lanvaux. Here a natural rock overhang has been converted into a closed tomb by the construction of a megalithic wall with returns running back to the rock face at either end (Gouézin 1994, 87-88). In some areas of Galicia, too, megalithic tombs were built up against natural rock outcrops (Criado Boado & Villoch Vázquez 2000). In all these instances, the boulders, outcrops or overhangs might have already been considered places of special significance, and the locations may have been chosen for that reason rather than simply from considerations of economy or efficiency.

Craggy uplands will often have provided accessible and sometimes pre-formed megalithic blocks. Thus scatters of fallen blocks surround the rocky outcrops of the Preseli Hills that were the source of the Stonehenge bluestones (Darvill, this volume). Accessible blocks would also have been available at coastal cliffs and headlands. Water-worn

---

[4] "on avait été forcé de choisir le lieu où la grosse pierre, destinée à être superposée, avait déjà sa place".

Stone heights

- 1.50 - 2 m
- 1.00 - 1.40 m
- 0.50 - 0.95 m
- <0.49 m
- stump

24m contour

23m

22m

21m

20m

19m

18m

N

0   50m

Upper slope

Central plateau

Lower slope

Large menhirs

Medium-sized menhirs

Small menhirs without weathering

a

Large blocks dominant

Medium blocks dominant

Small blocks (buried)

b

Widely spaced faulting

Intermediate

Closely-spaced faulting

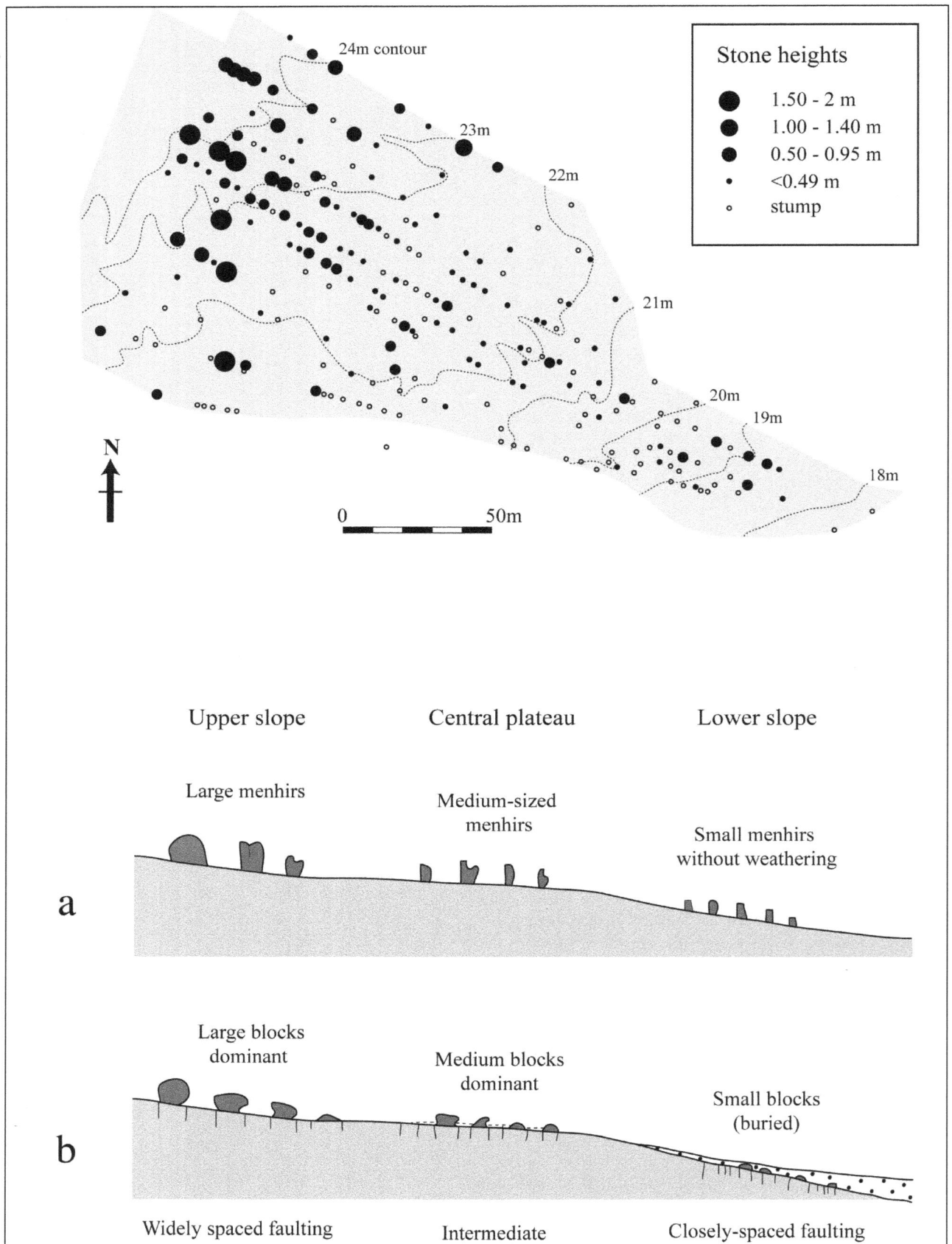

Fig. 1.4. Kerlescan alignments at Carnac (Morbihan): (above) plan of alignments indicating patterned distribution of stones of different dimensions; (below) (a) schematic cross-section through alignments, and (b) hypothetical reconstruction of the pre-monument landscape (after Sellier 1995)

Fig. 1.5. Carn Wnda, southwest Wales: megalithic tomb consisting simply of capstone raised at one end to create small space beneath (Photo: Chris Scarre)

stones have been recognised at a number of megalithic monuments, including La Hougue Bie on Jersey. Mourant noted that fifteen of the blocks from La Hougue Bie were visibly wave-worn, and others too were rounded. Only 15 of the 65 blocks were angular, and although these must have been "in a broad sense" quarried, Mourant adds that they too may have been obtained from rocks on the foreshore. The connection with the sea is reinforced by the quantities of shells found in the composition of the Hougue Bie cairn. These must have been specially brought from the shore, although the cairn itself was not built of shoreline material but local dolerite from the adjacent Queen's Valley (Mourant 1933).

The cairn of Barnenez, on a prominent headland on the north coast of Brittany, also incorporated wave-worn blocks in its construction. The principal building materials were the white and grey granites of Sterec, today an island but formerly linked to the Barnenez headland. The grey granite blocks have rounded contours and were probably collected directly from the foreshore. The white granite, by contrast, has smooth surfaces suggesting that it was quarried. A third material, pale Keriou granite, supplied some of the largest slabs in the monument but

these are thought not to have been quarried but to have been taken from the litter of blocks at the foot of an escarpment (Giot *et al.* 1995) (Fig. 1.6). Many of the megalithic slabs used in the Pornic passage graves are also thought to have come from coastal cliffs (Scarre 2004). At the Pointe du Souc'h, the southwestern extremity of Brittany, coastal cliffs provided the source of the megalithic blocks used in the tombs; but smaller granite blocks for the material of the cairns were obtained from quarries cut into the surface of the headland (Le Goffic, this volume).

In inland locations, river cliffs would have provided equally visible and accessible sources of megalithic blocks. Here, too, water-worn blocks are found, but in this case from river beds rather than coastal locations (Schierhold, this volume). Once again, the material for the tombs was derived mainly and perhaps exclusively from surface exposures, from sources that were visible in the landscape, and where blocks could be separated from the parent material with a minimum of effort. That said, however, the extraction of large blocks can never have been an easy undertaking, even where the outcrops or surfaces presented deep natural fissures. It would be

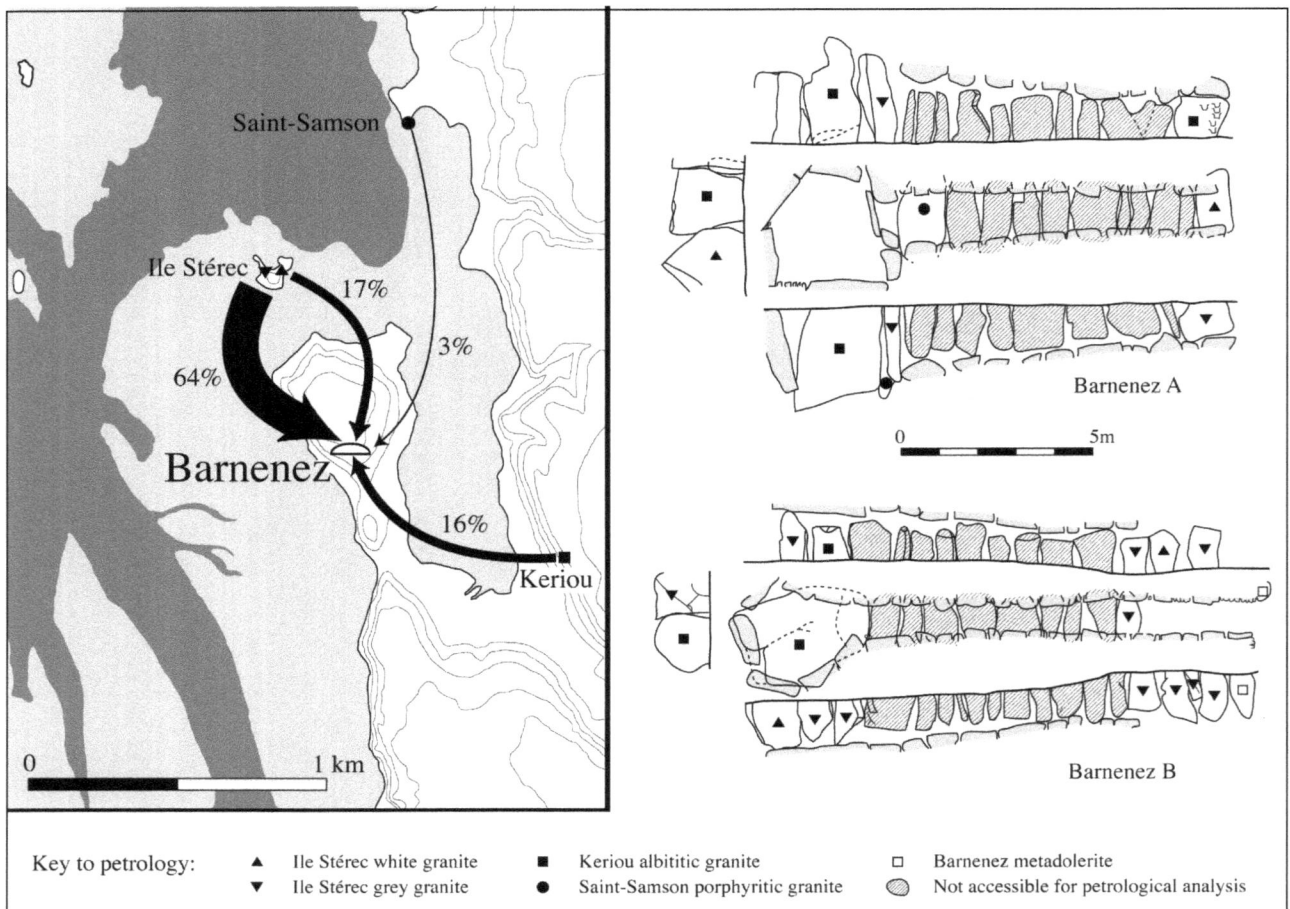

Fig. 1.6. Sources of megalithic blocks at Barnenez (Finistère): a) petrologies of individual slabs in passage graves A and B; b) sources and transport distances of all identified stone (after Giot *et al.* 1995)
Light grey indicates probable extent of Neolithic coastal lowland

misleading, furthermore, to conclude that megalithic blocks were always taken from the most convenient or accessible sources. The transport of the Stonehenge bluestones provides a striking demonstration of the power of sacred and mythological associations to drive Neolithic communities to bring material from distant and sometimes difficult locations. The same considerations would also have been important at the local scale, where we may envisage that certain sources or outcrops were more symbolically significant than others that may have been easier to exploit. These observations are equally relevant to the quarrying of blocks, where it may have been powers and associations of particular places as much as the presence of geologically suitable material that guided the location of extraction.

CUTTING AWAY: THE EVIDENCE FOR QUARRYING

It is clear from these examples that Neolithic communities may often have been able to collect megalithic blocks lying scattered on the surface of the land, or from jumbled masses of fallen blocks and scree at the base of cliffs or outcrops. Not all megalithic blocks were so easily obtained, however, nor need detached boulders always have been the preferred or symbolically most significant sources of material. Recent fieldwork has identified a growing list of quarry sites where specific traces of Neolithic working have been discovered. Several of them are described in the papers in this volume. To these can be added sites in Scotland and northern Britain, in northwest France, and in western Iberia.

In the far north of Britain, the quarrying of megalithic blocks has been identified at Vestra Fiold on Orkney. The size and shape of the monoliths was determined by the nature of the bedding planes in this sandstone outcrop, which makes them appear as if rising out of the ground. Excavation around one monolith showed that it had slipped off stone supports on which it had been suspended over a rock-cut pit (Fig. 1.7). The pit was probably dug to facilitate the mounting of the stone on a timber sledge. Petrologically the sandstone at Vestra Fiold is similar to that of two monoliths of the Stones of Stenness stone circle and five of those within the Ring of Brodgar. The varied lithologies of the stones in these two stone circles suggest that a whole series of megalithic quarries across

Fig. 1.7. Excavations at the megalithic quarry at Vestra Fiold (Orkney). (Photo: Colin Richards)

Orkney may have been worked to provide the materials (Richards 2004b). Thus, despite the availability of coastal cliffs, the builders of the Orkney circles preferred to cut away blocks that were projecting from the land.

The quarrying of rock outcrops in a rather different context has been revealed by excavations at Hunterheugh in Northumberland. Here a sandstone outcrop with cup and ring motifs was partially covered by a stone cairn containing a stone-slab cist that dates probably to the Early Bronze Age. Eight panels of rock art were revealed by excavation and surface clearance, and from their different degrees of erosion were divided into two chronological groups. Quarrying of the outcrop appears to have begun during the Neolithic period. Blocks with early cup and ring motifs had subsequently been removed for re-use elsewhere, and the surfaces exposed by these removals had themselves been carved with new, fresher cup and ring motifs (Waddington 2005). The incorporation of cup and ring marked blocks in megalithic monuments is a well-attested phenomenon although no Neolithic examples of this practice are recorded in the Hunterheugh area. On the opposite side of the Pennines, however, the large standing stone known as Long Meg in Cumbria has only a single

decorated surface and is thought to have been cut from a decorated cliff face. The red sandstone of Long Meg contrasts strikingly with the granite boulders of the adjacent stone circle known as her Daughters, and the difference in colour and lithology highlights the separate origin of Long Meg (Frodsham 1996, 111). A similar process is suggested by recent work at Torbhlaren in the Kilmartin valley of Mid-Argyll, where a decorated rock outcrop appears to have been quarried and may be the source of an adjacent decorated standing stone (Jones & O'Connor 2007).

Crossing the Channel, evidence has been found for the extraction of granite blocks of megalithic proportions on Ile Guennoc, a small island off the north coast of Brittany (Gouletquer 2000). They may represent abandoned Neolithic quarries, though they are unlikely to have been linked to the famous series of 5th millennium chambered cairns that line the low central ridge of Ile Guennoc since the latter are predominantly of dry-stone construction with corbelled vaults, and incorporate very few larger slabs. The scattered large blocks extracted from the two quarries on the northern edge of the island may instead have been intended for other kinds of megalithic monument, such as the *allée couverte* believed once to have existed on low ground that formerly linked Ile Guennoc to the mainland (Giot 1987, 137).

Detailed studies elsewhere in northwest France have revealed the manner in which outcrops were worked and megalithic blocks chosen and detached. At the Rocher Mouton, fire-setting was used to facilitate the extraction of monolithic slabs from a prominent granite outcrop (Mens, this volume). In the Vendée, blocks of naturally anthropomorphic shape were commonly chosen for the short rows of three to seven standing stones that cluster in the Avrillé area (Bénéteau, this volume). The preference for an anthropomorphic shape highlights the symbolic relationship of standing stones to people that has been widely discussed elsewhere (e.g. Parker Pearson & Ramilisonina 1998; Richards 2004b, 110; Scarre 2009). Bénéteau also observes how several of the Avrillé menhirs display cup marks which may have been present on the outcrops from which they were taken. The cup marks may have guided the choice of these particular blocks. Again this is part of a broader phenomenon with links to Hunterheugh, Long Meg and many other sites in Britain and other regions of western Europe. The cup-marked stone that forms the capstone of the central chamber of the Tumulus de Saint-Michel at Carnac in southern Brittany may be one of the earliest examples of this practice, and would indicate that the re-use of cup-marked slabs may already have begun by the middle of the 5th millennium BC in northwest France.

These examples demonstrate that even where megalithic blocks had to be cut away, the sources invariably lay accessible on or immediately below the surface of the ground. A study of weathering patterns on 2000 megalithic blocks in monuments of the Carnac area has revealed that 60% came from surface exposures (Mens, this volume). In Hesse and Westphalia, the limestone slabs for the tombs came from river cliffs and shallow deposits, in some cases from the beds of rivers themselves, exploiting the natural fracture patterns of the weathered parent material (Schierhold, this volume). The same practice is repeated in Portugal. The megalithic blocks used in the Vale de Rodrigo tombs in the Alentejo came probably from surface exposures. Vortisch has observed how the fracture planes of local granite outcrops in this region yield pre-formed monoliths of distinctive shapes that are suitable for the different parts of a megalithic chamber, and that could be detached from the parent outcrop with minimum effort. The small- and medium-sized capstones have a flat underside and a convex upper side. The weathering of the upper sides indicates that this was the original exposed surface of the bedrock. Larger, parallel-sided capstones may have been bedded horizontally in the source outcrop immediately below this uppermost weathered layer, and could also have been extracted by exploiting natural cleavage planes, especially where these were heavily weathered. The same observations can be extended to the orthostats. Their plano-convex or parallel-sided shape betrays their original

position within the outcrop (Vortisch 1999). In northern Portugal, at Cota near Viseu, it was surface material once again that was exploited for megalithic construction. A large exposed granite pavement 300m south of Cota 2 has a number of loose granite blocks corresponding in thickness to those in the tomb. As in the Alentejo, the perpendicular natural clefts and fissures produce blocks of a suitable size. In a study of these and other Portuguese tombs, only one, Fonte de Malga, had megalithic blocks with an unweathered character that showed they were not from surface deposits but must have come from a sub-surface quarry (Dehn et al. 1991).

Fonte de Malga is an exception, for it is clear that in the vast majority of cases, the megalithic blocks came from sources that were visible and accessible, in the form of cliffs, outcrops and boulder spreads. Thus the very concept of megalithic 'quarrying' must be qualified; only very rarely did it involve digging into the bedrock to any great depth. The blocks that were employed were not hidden or buried, but already detached or easily separated from their parent material. This observation has key significance for the whole purpose and meaning of 'megalithic' construction.

CONCLUSION

This brief review of megalithic stone sources and quarrying has emphasised the accessible and often visible nature of the places from which the blocks were taken. It would be wrong, however, to conclude that the process was essentially opportunistic in character, driven by considerations of least-effort and economy. Earlier writers pondered the question whether megalithic construction should be deemed a 'primitive' form of architecture, undertaken by societies technologically ill-equipped to break up the large slabs into smaller pieces. Fergusson rejected the notion of a progression from 'megalithic' to 'microlithic' architecture, arguing that the two 'styles' were entirely separate in origin and concept (Fergusson 1872, 42). The frequent use of dry-stone work alongside the megalithic blocks in chambered tombs is sufficient in itself to dispel any such lingering notion. Nor was the manipulation of megalithic blocks an easy option. The details of tomb construction revealed for instance by studies in Denmark (Dehn, this volume) illustrate the skill and complexity of the construction process. Furthermore, the geographical and chronological distribution of megalithic monuments in western and northern Europe argues strongly for a set of interlinked regional traditions with shared knowledge and expertise. And of course, we must not forget the immense variety of megalithic monuments, in which standing stones form a prominent part from the very outset. Standing stones are indeed probably the earliest megalithic forms in Brittany and Portugal. It is not impossible that the first chambered tombs consisted predominantly of re-used standing stones, some of them decorated, others not.

What is striking, and most significant, about the sources of the megalithic blocks is their intimate association with visible natural features. The massive slabs, largely unworked in many cases, would have provided an instant reminder of the landscape locations from which they had been taken. Some of them bore a close resemblance to natural formations, as we have seen in the example of the schist orthostats at Lajinha. The glacial erratics employed in north European tombs may likewise have recalled the boulder fields from which they came. In other cases, the original appearance of the blocks was masked by extensive smoothing or working of the surfaces, or was obscured by the application of plaster and painted decoration (notably at Dombate in Galicia: Bello Dieguez 1997). Yet even here the individual stones may have had special associations that they derived from their sources.

Ethnography and archaeology combine to suggest that the landscapes and landforms encountered by early farming societies would have been enculturated by the attribution of sacred and mythological significance to prominent features such as rivers, rocks, and trees. It is within such an enculturated landscape that the extraction of megalithic blocks for the Neolithic monuments of western Europe must be understood. The symbolic significance of stone will have formed part of a belief system that gave meaning not only to axe quarries and flint mines but must also have extended to megalithic blocks, and very probably the boulders, cliffs and outcrops from which they were taken. Ethnography provides relevant and indicative information about living or recent megalithic traditions, and highlights the key roles of animism and anthropomorphism in the beliefs held about stones and other natural materials by non-western societies. The proper study of megalithic monuments, if such can be defined, should hence combine attention to the materiality of the stones with examination of the methods by which and the places from which they were extracted, at the same time remaining ever mindful of the powerful symbolic dimensions.

Acknowledgements

I am grateful to Tim Darvill and Colin Richards for supplying photographs of Lockeridge and Vestra Fjold, and to Judith Roberts for preparing several of the figures. Thanks also to John Chapman and Emmanuel Mens for reading through and commenting on an earlier version of this paper.

References

ALLEN, M.J. (2005) Considering prehistoric environmental changes on the Marlborough Downs. In Brown, G.; Field, D.; Mcomish, D., eds. *The Avebury Landscape. Aspects of the field archaeology of the Marlborough Downs*. Oxford: Oxbow Books, 77-86.

BAKKER, J.A. (1992) *The Dutch Hunebedden. Megalithic Tombs of the Funnel Beaker Culture*. Ann Arbor, Michigan: International Monographs in Prehistory.

BAKKER, J.A.; GROENMAN-VAN WAATERINGE, W. (1988) Megaliths, soils and vegetation on the Drenthe Plateau. In Groenman-van Waateringe, W.; Robinson, M., eds. *Man-made Soils*. Oxford: British Archaeological Reports, 143-81.

BARBER, M.; FIELD, D.; TOPPING, P. (1999) *The Neolithic Flint Mines of England*. Swindon: English Heritage.

BARKER, C.T. (1992) *The Chambered Tombs of South-West Wales*. Oxford: Oxbow Books.

BELLO DIEGUEZ, J.M. (1997) Aportaciones del Dolmen de Dombate (Caban, La Coruña) al arte megalítico occidental. In L'Helgouach, J.; Le Roux, C.-T.; Lecornec, J. eds. *Art et Symboles du Mégalithisme Européen. Actes du 2ème Colloque International sur l'Art Mégalithique, Nantes 1995*. Nantes: Revue Archéologique de l'Ouest, supplément no.8, 23-39.

BOIVIN, N. (2004) From veneration to exploitation: human engagement with the mineral world. In Boivin, N.; Owoc, M.A., eds. *Soils, Stones and Symbols: cultural perceptions of the mineral world*. London: UCL Press, 1-29.

BONSTETTEN, Baron A. de (1865) *Essai sur les dolmens*. Geneva: Jules-Guillaume Fick.

BOUILLON, R. (1999) La sépulture mégalithique à entrée latérale du Petit Vieux-Sou à Brécé (Mayenne). *Revue Archéologique de l'Ouest* 6, 51-70.

BOWEN, H.C. & SMITH, I.F. (1977) Sarsen stones in Wessex: the Society's first investigations in the Evolution of the Landscape project. *Antiquaries Journal* 57, 185-96.

BOYER, P. (1996) What makes anthropomorphism natural: intuitive ontology and cultural representations. *Journal of the Royal Anthropological Institute* 2, 83-97.

BRUMM, A.; BOIVIN, N.; FULLAGAR, R. (2006) Signs of life: engraved stone artefacts from Neolithic south India. *Cambridge Archaeological Journal* 16, 165-90.

BUENO-RAMÍREZ, P.; BALBÍN BEHRMANN, R. de; BARROSO BERMEJO, R. (2007) Chronologie de l'art mégalithique ibérique: C[14] et contextes archéologiques. *L'Anthropologie* 111, 590-654.

BUKACH, D. (2003) Exploring identity and place: an analysis of the provenance of passage grave stones on Guernsey and Jersey in the Middle Neolithic *Oxford Journal of Archaeology* 22, 23-33.

BURTON, J. (1984) Quarrying in a tribal society. *World Archaeology* 16, 234-47.

CAULFIELD, S; O'DONNELL, R.G.; MITCHELL, P.I. (1998) ^{14}C dating of a Neolithic field system at Céide

Fields, County Mayo, Ireland. *Radiocarbon* 40, 629-40.

CHIPPINDALE, C. (1994) *Stonehenge Complete.* London: Thames & Hudson.

CLEAL, R.M.J.; WALKER, K.E.; MONTAGUE, R. (1995) *Stonehenge in its Landscape. Twentieth-century excavations.* London: English Heritage.

COONEY, G. (2000) *Landscapes of Neolithic Ireland.* London: Routledge.

COONEY, G. (2005) Stereo porphyry: quarrying and deposition on Lambay Island, Ireland. In Topping, P.; Lynott, M., eds. *The Cultural Landscape of Prehistoric Mines.* Oxford: Oxbow Books, 14-29.

CRIADO BOADO, F.; VILLOCH VÁZQUEZ, V. (2000) Monumentalizing landscape: from present perception to the past meaning of Galician megalithism (northwest Iberian peninsula). *European Journal of Archaeology* 3, 188-216.

CUMMINGS, V. (2002) All cultural things. Actual and conceptual monuments in the Neolithic of western Britain. In Scarre, C., ed., *Monuments and Landscape in Atlantic Europe.* London: Routledge, 107-21.

D'AULT-DUMESNIL, G. (1866) Recherches sur la provenance des granits qui ont servi à élever les monuments dits celtiques. *Bulletin de la Société Polymathique du Morbihan* 1866, 101-6.

DAIRE, M.-Y.; LE PAGE, G. (1994) Un monument mégalithique sur Enez Bihan en Plomeur-Bodou. *Bulletin de l'Association Manche-Atlantique pour la Recherche Archéologique dans les Iles* 7, 49-56.

DARVILL, T. (2006) *Stonehenge. The biography of a landscape.* Stroud: Tempus.

DE WISMES (1876) Le tumulus des Trois Squelettes à Pornic (Loire-Inférieure). *Bulletin de la Société Archéologique de Nantes* 15, 199–271.

DEANE, J.B. (1834) Observations on Dracontia. *Archaeologia* 25, 188-229.

DEHN, W.; KALB, P.; VORTISCH, W. (1991). Geologisch-Petrographische Untersuchungen an Megalithgräbern Portugals. *Madrider Mitteilungen* 32, 1-28.

ERIKSEN, P. (2004) Newgrange og den hvide mur. *Kuml* 2004, 45-77.

FERGUSSON, J. (1872) *Rude Stone Monuments in All Countries; their Ages and Uses.* London: John Murray.

FIELD, D. (2006) *Earthen long barrows: the earliest monuments in the British Isles.* Stroud: Tempus.

FREDERIC VII (1857) Sur la construction des salles dites des géants. *Mémoires de la Société Royale des Antiquaires du Nord.*

FRENCH, C.; LEWIS, H.; ALLEN, M.J.; SCAIFE, R.G.; GREEN, M. (2003) Archaeological and palaeo-environmental investigations of the upper Allen valley, Cranborne Chase, Dorset (1998-2000): a new model of earlier Holocene landscape development. *Proceedings of the Prehistoric Society* 69, 201-34.

FRODSHAM, P. (1996) Spirals in time: Morwick Mill and the spiral motif in the British Neolithic. *Northern Archaeology* 13/14, 101-38.

GEDDES, I.; WALKINGTON, H. (2005) The geological history of the Marlborough Downs. In Brown, G.; Field, D.; Mcomish, D., eds. *The Avebury Landscape. Aspects of the field archaeology of the Marlborough Downs.* Oxford: Oxbow Books, 58-65.

GELL, A. (1998) *Art and Agency. An anthropological theory.* Oxford: Clarendon Press.

GIOT, P.-R. (1987) *Barnenez, Carn, Guennoc.* Rennes: Travaux du Laboratoire d'Anthropologie, Préhistoire, Protohistoire et Quaternaire Armoricains.

GIOT, P.-R.; CHAURIS, L.; MORZADEC, H. (1995) L'apport de la pétrographie à l'archéologie préhistorique sur l'exemple du cairn de Barnenez en Plouezoc'h (Finistère). *Revue Archéologique de l'Ouest* 12, 171-6.

GOUÉZIN, P. (2004) *Les mégalithes du Morbihan intérieur des Landes de Lanvaux au Nord du département.* Rennes: Institut Culturel de Bretagne.

GOULETQUER, P. (2000) Fins de carrières à i'île Guennoc (Landéda, Finistère). In Cassen, S.; Boujot, C.; Vaquero, J., eds. *Eléments d'architecture. Exploration d'un tertre funéraire à Lannec er Gadouer (Erdeven, Morbihan). Constructions et reconstructions dans le Néolithique morbihannais. Propositions pour une lecture symbolique.* Chauvigny: Association des Publications Chauvinoises, 555-61.

GUTHRIE, S.E. (1993) *Faces in the Clouds: a new theory of religion.* New York & Oxford: Oxford University Press.

HERBERT, A. (1849) *Cyclops Christianus, or an Argument to disprove the supposed Antiquity of Stonehenge and other Megalithic Erections in England and Britanny.* London: J. Petheram.

JONES, A. (1999) Local colour: megalithic architecture and colour symbolism in Neolithic Britain. *Oxford Journal of Archaeology* 18, 339-50.

JONES, A.; O'CONNOR, B. (2007) Excavating art: recent excavations at the rock art sites at Torbhlaren, near Kilmartin, Mid-Argyll, Scotland. *Past* 57, 1-3.

KALB, P. (1996) Megalith-building, stone transport and territorial markers: evidence from the Vale de Rodrigo, Évora, south Portugal. *Antiquity* 70, 683-5.

KENDRICK, T.D. (1925) *The Axe Age. A study in British prehistory.* London: Methuen & Co,.

KING, V.T. (1976) Stones and the Maloh of Indonesian West Borneo. *Journal of the Malaysian Branch of the Royal Asiatic Society* 48, 104-19.

L'HELGOUACH, J. (1983) Les idoles qu'on abat ... (ou les vicissitudes des grandes stèles de Locmariaquer).

Bulletin de la Société Polymatique du Morbihan 110, 57-68.

L'HELGOUACH, J. (1997) De la lumière au ténèbres. In In L'Helgouach, J.; Le Roux, C.-T., eds. *Art et Symboles du Mégalithisme Européen. Actes du 2ème Colloque International sur l'Art Mégalithique, Nantes 1995.* Nantes: Revue Archéologique de l'Ouest, supplément no.8, 107-23.

LARSSON, L. (1998) Rock, stone and mentality. Stones that unite, stones that subjugate – a megalithic tomb in the Vale de Rodrigo, southern Portugal. *KVHAA Konferenser* 40, 137-55.

LE ROUX, C.-T.; LECERF, Y.; GAUTIER, M. (1989) Les mégalithes de Saint-Just (Ille-et-Vilaine) et la fouille des alignements du Moulin de Cojou. *Revue Archéologique de l'Ouest* 6, 5-29.

LISLE DU DRÉNEUC, P. de (1882) *Dictionnaire Archéologique de la Loire-Inférieure (Epoques primitive, celtique, gauloise et gallo-romaine).* Nantes.

LUKIS, F.C. (1853) Observations on the Celtic megaliths, and the contents of Celtic tombs, chiefly as they remain in the Channel Islands. *Archaeologia* 35, 232-58.

LYNCH, F. (1975) Excavations at Carreg Samson megalithic tomb, Mathry, Pembrokeshire. *Archaeologia Cambrensis* 124, 15-35.

MARGUERIE, D. (1992) *Evolution de la végétation sous l'impact humain en Armorique du Néolithique aux périodes historiques.* Rennes: Laboratoire d'Anthropologie, Préhistoire, Protohistoire et Quaternaire Armoricains.

MARGUERIE, D. (2000) Végétation néolithique sous impact anthropique en Morbihan et dans le reste de la Bretagne. In Cassen, S.; Boujot, C.; Vaquero, J., eds. *Eléments d'architecture. Exploration d'un tertre funéraire à Lannec er Gadouer (Erdeven, Morbihan). Constructions et reconstructions dans le Néolithique morbihannais. Propositions pour une lecture symbolique.* Chauvigny: Association des Publications Chauvinoises, 563-6.

MEIGHAN, I.G.; SIMPSON, D.D.A.; HARTWELL, B.N. (2002) Newgrange – sourcing of its granitic cobbles. *Archaeology Ireland* 16, 32-5.

MEIGHAN, I.G.; SIMPSON, D.D.A.; HARTWELL, B.N.; FALLICK, A.E; KENNAN, P.S. (2003) Sourcing the quartz at Newgrange, Brú na Bóinne, Ireland. In Burenhult, G., ed. *Stones and Bones. Formal disposal of the dead in Atlantic Europe during the Mesolithic-Neolithic interface 6000-3000 BC.* Oxford: BAR Publishing, 247-51.

MENS, E. (2007) Étude technologique des mégalithes de l'Ouest de la France, les monuments néolithiques du Mané-Bras et du Mané-Bihan à Locoal-Mendon (Morbihan). In Évin, J., ed. *Un siècle de construction du discours scientifique en Préhistoire, 3: Aux conceptions d'aujourd'hui.* Paris: Société Préhistorique Française, 353-9.

MOHEN, J.-P.; SCARRE, C. (2002) *Les Tumulus de Bougon. Complexe mégalithique du Ve au IIIe millénaire.* Paris: Errance.

MOURANT, A.E. (1933) Dolmen de la Hougue Bie. Nature and provenance of materials. *Bulletin Annuel of the Société Jersiaise* 12, 217-20.

MOURANT, A.E. (1939) Notes on petrology. In Hawkes, J., ed. *The Archaeology of the Channel Islands. Volume II: the Bailiwick of Jersey.* Saint Helier: Société Jersiaise, 205-7, 12-3, 25, 38-9, 44-5, 69.

MOURANT, A.E. (1963) The stones of the Mont de la Ville passage grave, Jersey. *Annual Bulletin of the Société Jersiaise* 88, 317-28.

MOURANT, A.E. (1977) The use of Fort Regent granite in megalithic monuments in Jersey. *Annual Bulletin of the Société Jersiaise* 22, 41-9.

O'SULLIVAN, M. (2009) Living Stones – Geology and Ideology in the Neolithic. In Cooney, G.; O'Connor, B.; Chapman, J., eds. *Materialitas: working stone, carving identity.* Prehistoric Society Research Papers 1

PARKER PEARSON, M.; RAMILISONINA (1998) Stonehenge for the ancestors: the stones pass on the message. *Antiquity* 72, 308-26.

PATTON, M. (1992) Megalithic transport and territorial markers: evidence from the Channel Islands. *Antiquity* 66, 392-5.

PATTON, M.; RODWELL, W.; FINCH, O. (1999) *La Hougue Bie, Jersey. A study of the Neolithic tomb, medieval chapel and Prince's Tower, including a report in the excavations of 1991-95.* St Helier: Société Jersiaise.

PEET, T.E. (1912) *Rough Stone Monuments and their Builders.* London: Harper & Brothers.

PIGGOTT, S. (1985) *William Stukeley. An Eighteenth-Century Antiquary.* London: Thames & Hudson.

POLLARD, J.; REYNOLDS, A. (2002) *Avebury: The biography of a landscape.* Stroud: Tempus.

POMERAI, M.R. de (1997) Neolithic engineering: the use of stone in Guernsey passage graves. *Transactions of the Guernsey Society of Natural Science and Local Research* 24, 284-9.

RICHARDS, C. (1996) Monuments as landscape: creating the centre of the world in Neolithic Orkney. *World Archaeology* 28, 190-208.

RICHARDS, C. (2004a) Labouring with monuments: constructing the dolmen at Carreg Samson, south-west Wales. In Cummings, V.; Fowler, C., eds. *The Neolithic of the Irish Sea. Materiality and traditions of practice.* Oxford: Oxbow Books, 72-80.

RICHARDS, C. (2004b) A choreography of construction: monuments, mobilization and social organization in Neolithic Orkney. In Cherry, J.; Scarre, C.; Shennan, S., eds. *Explaining Social Change: studies in honour of Colin Renfrew*. Cambridge: McDonald Institute for Archaeological Research, 103-13.

SCARRE, C. (2002) A place of special meaning: interpreting prehistoric monuments through landscape, in *Inscribed Landscapes: marking and making place*, eds. B. David & M. Wilson. Honolulu: University of Hawaii Press, 154-75.

SCARRE, C. (2004) Choosing stones, remembering places: geology and intention in the megalithic monuments of western Europe. In Boivin, N.; Owoc, M.A., eds. *Soils, Stones and Symbols. Cultural perceptions of the mineral world*. London: UCL Press, 187-202.

SCARRE, C. (2009) Stones with character: ani-mism, agency and megalithic monuments. In Cooney, G.; O'Connor, B.; Chapman, J., eds. *Materialitas: working stone, carving identity*. Prehistoric Society Research Papers 1

SCHULDT, E. (1972) *Die mecklenburgischen Megalithgräber. Untersuchungen zur ihrer Architektur und Funktion*. Berlin: Deutscher Verlag der Wissenschaften.

SCOTT, D.D.; THIESSEN, T.D. (2005) Catlinite extraction at Pipestone National Monument, Minnesota: social and technological implications. In Topping, P.; Lynott, M., eds. *The Cultural Landscape of Prehistoric Mines*. Oxford: Oxbow Books, 140-54.

SELLIER, D. (1991) Analyse morphologique des marques de la météorisation des granites à partir de mégalithes morbihannais. L'exemple de l'alignement de Kerlescan à Carnac. *Revue Archéologique de l'Ouest* 8, 83-97.

SELLIER, D. (1995) Eléments de reconstitution du paysage prémégalithique sur le site des alignements de Kerlescan (Carnac, Morbihan) à partir de critères géomorphologiques. *Revue Archéologique de l'Ouest* 12, 21-41.

SPROCKHOFF, E. (1938) *Die nordische Megalithkultur*. Berlin & Leipzig: Walter de Gruyter.

SUTTON, K. (1977) Reclamation of wasteland during the eighteenth and nineteenth centuries. In Clout, H.D., ed. *Themes in the Historical Geography of France*. London: Academic Press, 247-300.

THOMAS, H.H. (1923) The source of the stones of Stonehenge. *Antiquaries Journal* 3, 239-60.

THORPE, R.S.; WILLIAMS-THORPE, O. (1991) The myth of long-distance megalithic transport. *Antiquity* 65, 64-73.

TREVARTHEN, D. (2000) Illuminating the monuments: observation and speculation on the structure and function of the cairns at Balnuaran of Clava. *Cambridge Archaeological Journal* 10, 295-315.

VAN GIFFEN, A.E. (1927) *The Hunebeds in the Netherlands*. Utrecht: A. Oosthoek.

VORTISCH, W. (1999) Geologisch-petrographische Untersuchungen an megalithischen Monumenten – Beispiele aus Portugal. In Beinhauer, K.W.; Cooney, G.; Guksch, C.E.; Kus, S., eds. *Studien zur Megalithik. Forschungsstand und ethnoarchäologische Perspektiven*. Weissbach: Beier & Beran, 275-88.

WADDINGTON, C. (2005) Excavation of a rock art site at Hunterheugh Crag, Northumberland. *Archaeologia Aeliana* 34, 29-54.

WHITTLE, A. (1993) The Neolithic of the Avebury area: sequence, environment, settlement and monuments. *Oxford Journal of Archaeology* 12, 29-53.

WHITTLE, A. (2004) Stones that float to the sky: portal dolmens and their landscapes of memory and myth. In Cummings, V.; Fowler, C., eds. *The Neolithic of the Irish Sea. Materiality and traditions of practice*. Oxford: Oxbow Books, 81-90.

THE MEGALITHIC BUILDING SITE

Torben DEHN

Heritage Agency of Denmark, Ancient Monuments, H.C. Andersens Boulevard 2, DK 1553 Copenhagen K

Abstract: In this paper two aspects of Danish megalithic architecture are discussed: incorporated stones, and secondary modifications to monuments in prehistoric times. One example of the 'incorporated stone' is a closing stone, marking the position where a channel giving access into the chamber was left open during the building process. Examples of secondary modifications are the conversion of a dolmen into a passage grave, or the evidence for accidents that occurred during the building process. This evidence provides the basis for a new interpretation of megalithic architecture.
Key-words: megalith, passage grave, construction processes, architecture

Résumé: Dans cette étude deux aspects de l'architecture mégalithique danoise sont examinés : d'une part, "les pierres introduites", et d'autre part, les modifications secondaires qu'ont subies ces monuments pendant les périodes préhistoriques. Un exemple de 'pierre introduite', est la pierre de fermeture, qui marque la position dans laquelle un couloir d'accès a été laissé ouvert durant la construction. Sous le terme de 'modifications secondaires' on peut citer la transformation d'un dolmen en dolmen à couloir, ou encore les indices d'accidents de travail survenus pendant la construction. Ces observations nous fournissent les bases d'une nouvelle lecture de l'architecture mégalithique.
Mots-clés: mégalithe, dolmen à couloir, méthodes de construction, architecture

In studies of megalithic architecture the manner in which the large stones were handled is often discussed. This topic connects closely with the process of construction itself, and the traces of the building methods that appear in the structure of the chamber and within the surrounding mound. This paper deals with two aspects of this evidence: the use of 'incorporated stones', and the secondary modifications that some of these megalithic structures underwent. The starting point is provided by observations of the directly visible parts of these structures. These are combined with archaeological investigation, the latter involving both excavation and archival research since crucial information is often found in older descriptions of these sites.

INCORPORATED STONES

The term 'incorporated stones' refers to slabs which seem to be incorporated within the megalithic structure for reasons which are not only functional but may also have been artistic or religious. These 'incorporated stones', when considered as a separate category, provide a good opportunity to gain a better understanding of the concepts that lay behind the monuments and the thoughts of the people who built them.

One such concept is the phenomenon of dualism and symmetry, which has been examined in a number of recent studies (Dehn & Hansen 2000; 2006). Its physical manifestation takes various forms including twin stones, twin passage graves built together, pairs of separate but twin-like graves, twin-like passages leading to a single symmetrical chamber, and even two chambers symmetrical to each other. These architectural expressions of dualism clearly reflect a crucial principle in the world order of Neolithic societies. They are examples of the concepts that lay behind megalithic structures.

Other examples of incorporated stones can be observed in particular in the twin passage graves, where the symmetry of the two chambers makes any differences between them all the more distinct. An example is provided by the twin passage grave at Hvalshøje (or Iglsø), where two graves – apparently placed symmetrically within the Neolithic mound – are identical to each other except for their roofs (Dehn & Hansen 2006a, 31-32). The left-hand chamber is covered by two horizontal capstones, which is the customary arrangement in this region. The roof of the right-hand chamber, however, consists of three capstones stacked up against each other in a very unusual way. At first sight one might explain this by the lack of suitable material, but that does not seem probable since there appear to have been many suitable stones in the region. Furthermore, this passage grave is the earliest of eight passage graves that have been dated by birch bark from the dry-stone walling; it is unlikely that suitable local material had been used up at such an early stage in the sequence of megalith-building.

An incorporated element of special importance is the closing stone. These closing stones are often narrow and carefully shaped, with edges almost like a wedge. To some degree they are a functional feature, but a religious significance cannot be excluded. The closing stone takes the form of a narrow orthostat the base of which is often set a short distance above floor level. These orthostats are not integrated in the load-bearing structure but were obviously added as one of the final steps in the building process. An example is provided by the passage grave of Birkehøj, where excavation showed that an access corridor between the narrow end of the chamber and the area beyond the mound remained open during the entire building process (Dehn *et al.* 2004, 165-166). Finally the closing stone was put into position from behind, fitting between two orthostats like a cork in a bottle, and the mound material was added. The opening was used for

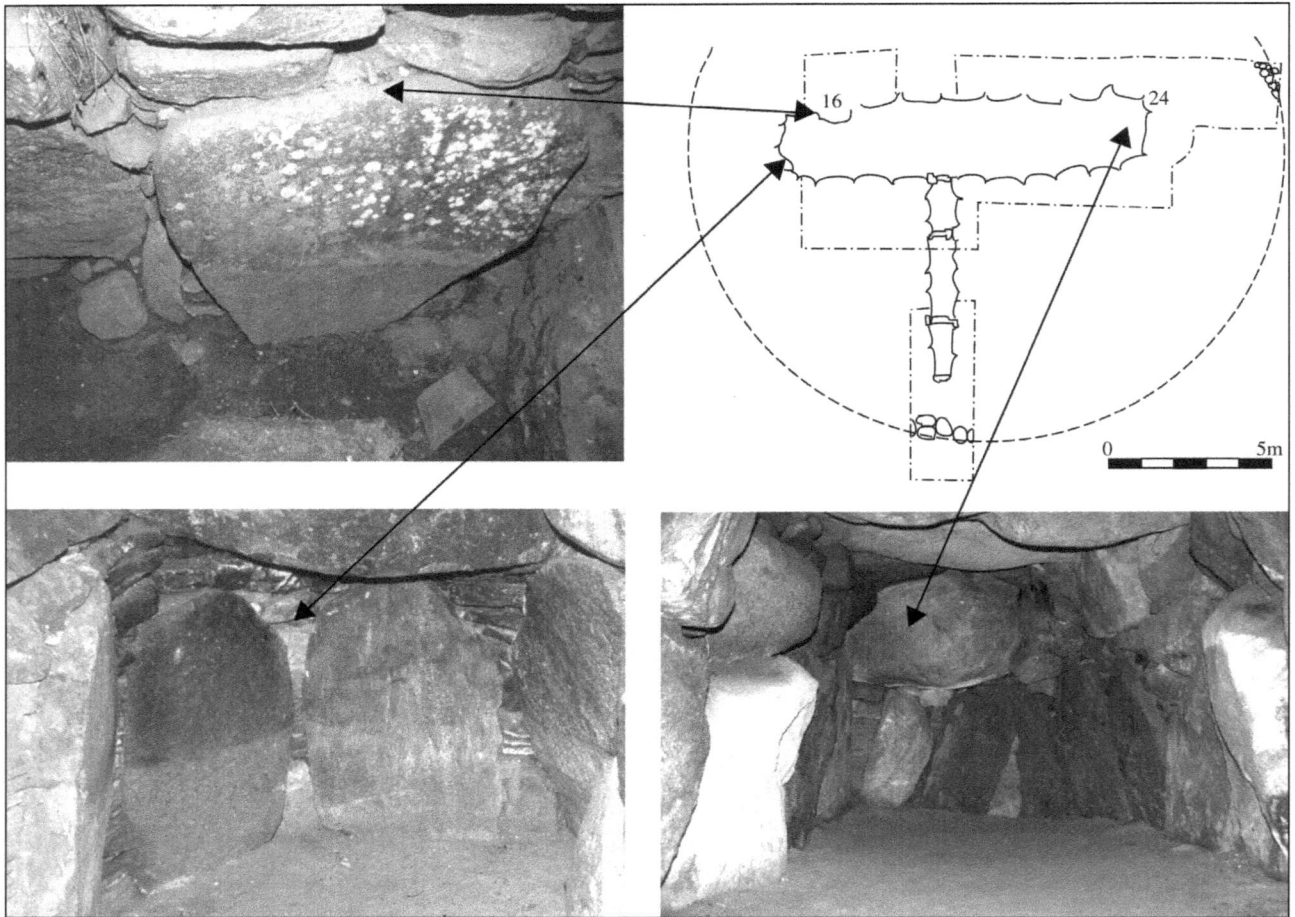

Fig. 2.1. Birkehøj passage grave. The narrower western end of the chamber (at left of ground plan, top right) was originally part of a smaller dolmen chamber that was left in original position (below left). The construction method here is distinctively different from the chamber of the passage grave that was added later (below right). The orthostat (above left) is a reused and relocated orthostat from the dolmen. It seems that shortly after its erection this orthostat toppled and had to be re-erected, so that it is now standing in incorrectly and the dry-stone walling around it has clearly been built up from inside the chamber. The capstones are not shown on the ground plan. (Drawing and photo: Torben Dehn & Jørgen Westphal)

transportation of materials; in fact at the end of the building process some surplus material from the final regulation or adjustment of the chamber's floor was deposited in the material at floor level was removed at the end of the building process and deposited in the access corridor just behind the closing slab.

It is possible that the provisional access corridor was made for other purposes too. At some time during the construction process the status of the building changed from that of a building site to that of a sacred monument. The main passage to the tomb that served to connect the dead and the living had up to three door slabs and must have been an important feature. One may imagine, however, that a point was reached when this passage was no longer available for profane use, and that could be another reason for the access corridor at the narrow end of the tomb chamber. Of course the transformation of the tomb into a sacred place might have been achieved through a ceremony of consecration when the monument

was totally completed, but a gradual process of consecration during the course of the building operation is an alternative worth considering. Viewed in that perspective it is the secondary entrance that is the interesting element, not the orthostat itself.

The classic example of incorporated stones is provided by the broken menhirs in the tombs of Locmariaquer. A corresponding example in a Danish megalith is found in the already mentioned passage grave of Birkehøj, which is unusual because of its size and challenging form. It could in fact be an expanded dolmen or a dolmen that has been incorporated in a passage grave. Unfortunately there is only architectural evidence, not archaeological evidence, for such a transformation but if one is familiar with megalithic structures built of erratic boulders the conclusion becomes fairly obvious (Fig. 2.1). The uprights and dry-stone walling at the narrower western end of the 11.5 metre-long chamber have a totally different character to that of the orthostats in the rest of the chamber. The uprights are smaller, with flat inner sides, and the dry-stone walling is well-built consisting of thin small flagstones. In the rest of the chamber the uprights are bigger with convex inner sides and the dry-stone walling is made of large clumsy stones (Dehn *et al.* 2004, 168-171).

A possible explanation for these differences is that the capstone and two uprights from a circular dolmen were removed and re-used in other positions within the passage grave, while the intact part of the dolmen today forms the narrow western end of the passage grave chamber. The incorporation of the stones removed from the dolmen occasioned severe problems for the megalith builders. The size, form and thickness of the boulders of the dolmen were different from the 'new' ones, and their technical integration presented a challenge. Another problem was the foundations. The ground beneath those uprights of the dolmen that were removed proved unstable and created difficult conditions which resulted later in a collapse.

These are only a few examples of the incorporation of non-functional elements in Danish megaliths. Within the European context the incorporation of such elements is not unusual but in the Scandinavian context this is a relatively new discovery.

It is interesting to note that these incorporated elements were not visible under normal circumstances, since the passage giving access to the chamber was probably carefully closed and would only have been opened on special occasions. Yet even though they normally were not visible these incorporated elements – or the knowledge or myths associated with them – seem to have been important. Those who participated in the building process knew about these details and their purpose, and were able to discuss and compare them to similar details in other monuments. Stories may also have circulated about occurrences during the building process: and the myths may have survived for generations. The shared experience which this represented may have been important for the strength and solidarity of society.

SECONDARY MODIFICATIONS

Even when the construction process was finished and the consecration ceremonies were completed, megalithic structures continued to be altered over time. In the course of excavation it can often be very difficult to date constructional changes and to identify the events that lay behind later modifications. Such modifications might be made immediately after the building was completed, during subsequent periods of prehistory, or within the last 200 years! For the building process itself it is the examples where Neolithic communities seem to have been involved that are the most interesting.

One type of modification is the consequence of mistakes made during the building process. In modern terminology

we refer to 'shoddy' construction work, and although megalithic monuments generally are characterized by precision, care and unbelievable capability this phenomenon is sometimes encountered in a megalithic context. At Birkehøj the incorporation of the upright from the dolmen resulted in a collapse because of the unstable ground. From the character of the repair this event can be judged to have occurred during the Neolithic period. An orthostat fell, but without displacing the capstones. The structure was repaired, but to the Neolithic people involved it must have been a nightmare, standing in the dark chamber to re-erect the heavy stone, and trying to put back behind the re-erected stone the pebble-fill that had spilled into the chamber. The result was a kind of emergency solution: the orthostat is in the wrong position, and the dry-stone walling has been built up from inside the chamber. But the event will have provided material for myths for many generations. The same technical problem was encountered during the recent restoration of the monument, and even with modern equipment and without the capstones in position it presented a severe challenge, serving as an instance of experimental archaeology (Dehn *et al.* 2004, 169-171).

The collapse at Birkehøj presumably occurred shortly after the monument was finished. Two other examples illustrate accidents during the process of placing the capstones. In the passage grave of Flintinge Byskov a panel of dry-stone walling in the corner of the chamber drew attention because it was twisted, a detail that had never been seen before (Dehn & Hansen 2006, 51-52) (Fig. 2.2). The upper part of the dry-stone walling filled a concavity in the face of the orthostat. This concavity was the result of a crack that had been caused by the sudden heavy weight of a capstone over which the builders lost control as they were lifting it onto the three orthostats. One of the orthostats cracked at the top, a piece of it fell away and the dry-stone walling of carefully shaped flagstones collapsed. The capstone was not removed, however, and its position was not corrected. This interpretation is confirmed by observation of certain details on the tops of the two other orthostats.

The sequence of stage is complicated. When the capstone was placed in position, resting on orthostats at two or three points, the seating for it was prepared by thin layers of flagstones that were carefully designed to ensure the correct angle of rest of the capstone. Once the capstone was in place, its position was further secured by adding wedge-shaped flagstones from the outside. Close study of these primary and secondary flagstones confirmed the supposition that an unfortunate accident had occurred, since some of the primary flagstones were cracked. Therefore, it is probable that the present dry stone walling is a secondary repair, adapted to the shape of the orthostat that was damaged by the weight of the capstone as it slipped out of control.

Another example illustrates the same phenomenon. One capstone inside the chamber of Jordehøj seems to be

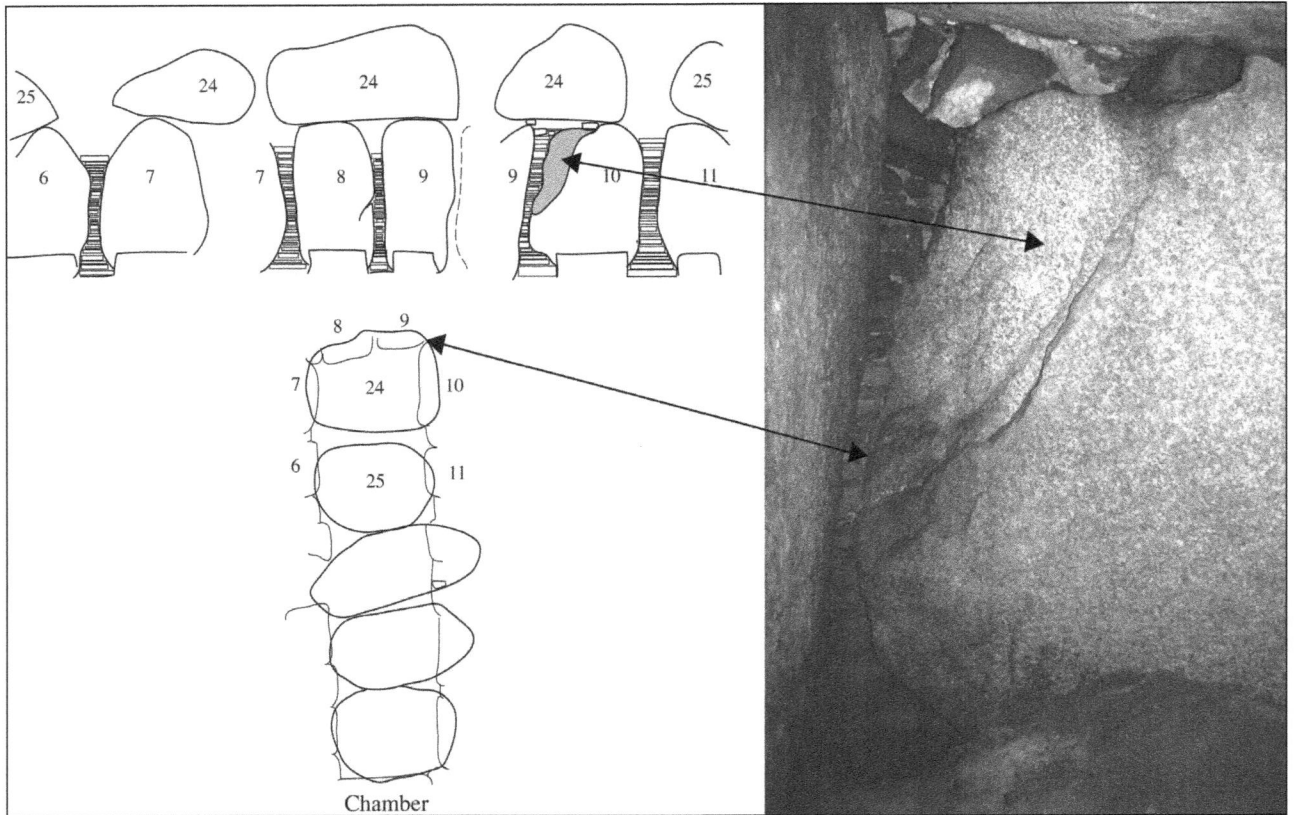

Fig. 2.2. Flintinge Byskov passage grave. The fracture on orthostat 10 is the result of cracking caused by the sudden heavy weight from a capstone out of control when it was lifted upon the three orthostats. Some of the flagstones on the top of the orthostats were broken in the process, and the dry-stone walling was partly destroyed by the accident and had to be repaired with new walling that was twisted in arrangement.
(Drawing and photo: Torben Dehn)

mislocated. The underside of the stone slopes so much that one of its ends passes in front of the upper part of the orthostats (Dehn *et al.* 2000, 92-104). This situation is exceptional. Slightly inclined capstones lying on wall stones of different heights are often seen, as are single low-lying capstones dividing a chamber into two sections, but the situation at Jordehøj is quite abnormal. Ever since the chamber was opened in 1836 this situation has been regarded as the result of degradation, but during an investigation in 1988 the upper side of the capstone was exposed, showing that its present position is the original one. Within the mound the two layers of flagstones covering the capstone were intact, indicating that the unusual position of the stone was accepted before the monument was finished.

Flintinge Byskov and Jordehøj illustrate situations where the positioning of a capstone was not properly controlled and where the consequent mistake was not corrected. An irregularly placed capstone is a particularly striking element in the stringent architecture that characterizes the Scandinavian megaliths. Working with megalithic structures makes one aware that the monuments can vary considerably among themselves but that this variability is constrained within a certain framework, a framework

defined by care, precision, system and regularity. One may therefore wonder why capstones that ended up in the wrong position were not corrected. A number of large orthostats, too, seem to stand in a peculiar manner, but it is not certain that mistakes were the cause. Was it technically impossible, or were there irrational reasons of some kind for not correcting the mistakes?

The stones in question are of average weight (around 10 tonnes) and of dimensions that do not normally appear to have been problematic. At this late stage in the constructional process when the capstones were being positioned, however, the whole structure of orthostats, dry-stone walling and the packing materials behind the walls risked being disturbed by the uncontrolled movement of a falling capstone. It may therefore have been judged better simply to stabilize the situation. It is known from experience during restoration work that even small pressure on an orthostat from a stone suspended from a crane must be avoided. The fear of worsening the situation may hence be the explanation, but the possibility that spiritual forces of some kind were involved in the construction work cannot be excluded. From ethnographic descriptions it is clear that complicated rules might have governed the process of collecting materials and building the tombs. This must be kept in mind when such irregularities in tomb structure are explained.

Any discussion of secondary modifications must include mention of Late Neolithic and Bronze Age interventions, since the latter serve to emphasize the supreme technique of the megalith builders. Late Neolithic insertions of new burials are a relatively common phenomenon and were not always achieved by opening the door slabs in the passage. In some cases they broke through the upper part of the end wall of the chamber. At Maglehøj a hole was dug into the mound, through the packing and through the material covering the roof, and an opening made between the capstone and the top of one of the orthostats (Dehn & Hansen 2007, 18-20). The reinstatement of the tomb after the burial had been inserted is characterized by a simple technique: a number of the largest flagstones were placed upright to close the gap in the chamber wall and the rest of the materials that had been dug out were just thrown randomly into the hole. The original megalith builders would have restored the tomb with flagstones laid in horizontal courses, and would then have re-established the roof covering in order to keep the construction tight and dry. It is remarkable that this careful building technique was no longer used even a few centuries later. It is not only a matter of forgotten technique. Late Neolithic people handled the same materials – stone, clay, timber – in daily life, and would have been able to build more carefully and solidly had they wanted to, albeit in not such a sophisticated manner as their Middle Neolithic forebears. They simply did not intend to build durable tombs because the idea behind burial, their idea of life and death, was quite different.

Examples of Bronze Age intrusions are also known. A similar principle was used in two passage graves in different regions of Denmark. At one of them, Stuehøj in Ølstykke,[1] an end stone was exposed from behind, was shattered by fire and was then removed in fragments. Two boulders from the structure were erected in the middle of the chamber and an oak coffin was probably placed between them. The end wall was rebuilt with a pile of boulders and smaller stones taken from the megalithic structure, and the surrounding mound was converted into a Bronze Age mound by adding more earth – a common practice.

Here again the intention was obviously very different from that of the original Middle Neolithic tomb-builders. Bronze Age oak coffins are normally buried in turf mounds; there is no evidence of any ability to handle large stones. Considerable energy was used to shatter the orthostat with fire instead of re-using it to close the grave.

It seems that familiarity with handling large stones had been lost by the Bronze Age and the crucial aspect of the monument was its external appearance, the content being the person buried inside the tomb. For megalithic monuments appearance was important too, but great importance was also attached to the details of internal construction. Such later intrusions and clumsy repairs confirm the conclusion that the Middle Neolithic megalith builders represent one of the high points in the history of building techniques.

Acknowledgements

The investigations which are the basis for this article have been made in collaboration with Svend Illum Hansen and Jørgen Westphal.

References

DEHN, T.; HANSEN, S.I. (2000) Doubleness in the Construction of Danish Passage Graves. In Ritchie, A., ed. *Neolithic Orkney in its European context.* Cambridge: McDonald Institute for Archaeological Research, 215-221.

DEHN, T.; HANSEN, S.I. (2006) Megalithic architecture in Scandinavia. In Laporte, L., Joussaume, R., & Scarre, C. eds., *Origin and development of the megalithic monuments of Western Europe.* Bougon: Musée de Bougon, 39-61.

DEHN, T.; HANSEN, S.I. (2006a) Birch bark in Danish passage graves. *Journal of Danish Archaeology* 14, 23-44.

DEHN, T.; HANSEN, S.I. (2007) Examples of megalithic technology and architecture in Denmark. In Bloemers, J.H.F. ed., *Tussen D26 en P14: Jan Albert Bakker 65 jaar.* Amsterdam: Amsterdams Archeologisch Centrum, Universiteit van Amsterdam, 17-31.

DEHN, T.; HANSEN, S.I.; KAUL, F. (2000) – *Klekkendehøj og Jordehøj. Restaureringer og undersøgelser 1985-90. Stenaldergrave i Danmark 2.* Copenhagen: Nationalmuseet, Skov- og Naturstyrelsen.

DEHN, T.; HANSEN, S.I.; WESTPHAL, J. (2004) Jættestuen Birkehøj. Restaureringen af en 5.000 år gammel storstengrav, *Nationalmuseets Arbejdsmark* 2004, 153-173.

---

[1] Unpublished investigation 2006.

HUNEBEDDEN AND HÜNENGRÄBER: THE CONSTRUCTION OF MEGALITHIC TOMBS WEST OF THE RIVER ELBE

Jan Albert BAKKER

Bothalaan 1, NL-3743 CS Baarn, The Netherlands

Abstract: The megalithic tombs of the TRB West Group were built from erratic boulders between 3400 and 3000 cal BC. It is not now possible to identify the places from which the builders took these boulders. The largest capstones weighed 40-60 tons, but usually they were much smaller. The large boulders were probably transported on sledges, and the capstones dragged into position up a wooden ramp, as illustrated by present practice on Sumba in Indonesia. A drawing from 1809 shows one small tomb completely covered by a mound. This tomb may have had a capping of slabs and loam to divert rainwater, as is found in the TRB North Group. The large tombs, however, and several of the smaller tombs, had mounds that barely covered the lower edges of capstones and the small stones in the voids between them in the roof of the chamber. This leaves us with the question: were these chambers kept dry, or were they not?

Key-words: hunebeds, glacial erratics, Netherlands, TRB

Résumé: Les sépultures mégalithiques du groupe occidental de la civilisation des vases en entonnoir (TRB) ont été construites avec des blocs erratiques entre 3400 et 3000 avant J.-C. Il n'est plus possible d'identifier la provenance des pierres utilisées par les bâtisseurs. Les dalles de couvertures les plus grandes pouvaient peser jusqu'à 40-60 tonnes, mais la plupart d'entre-elles étaient beaucoup moins grandes. Les blocs les plus grands ont probablement été déplacés sur des traîneaux, et les pierres de couverture manœuvrées et positionnées à l'aide d'une rampe, comme en témoignent les pratiques actuelles sur l'île de Sumba en Indonésie. Un dessin de 1809 représente une petite sépulture entièrement recouverte par son tumulus. Cette sépulture avait peut-être une couverture de dalles et de terre pour détourner l'eau de pluie, tout comme on le retrouve dans le groupe septentrional de la civilisation des vases en entonnoir. Pour les grandes sépultures cependant, et pour quelques unes des sépultures les plus petites, les tumulus montaient à peine jusqu'au bord inférieur des dalles de couvertes et de la pierraille qui comblait les vides entre les dalles de couverture de la chambre. Ce qui nous pose la question suivante: ces chambres se sont-elles maintenues à sec?

Mots-clés: hunebeds, blocs erratiques, Pays-Bas, civilisation des vases en entonnoir

The megalithic tombs of the northeastern Netherlands, known as *hunebedden*, and those of northwest Germany which are called *Grosssteingräber* or *Hünengräber*, have a central framework of large erratic boulders (Fig. 3.1). These boulders were transported to these areas by the glaciers of the Ice Age, especially during the Saalian glaciation. They derive from Finland, Scandinavia and the Baltic, and many of them lay scattered across the surface of the local sands of the Drenthe Plateau in the period 3400-3000 cal BC, when the tombs were built. The glaciers extended from the North Sea across the whole North European (or German) Plain, as far as the hills or low mountains of the *Mittelgebirge*. Their limits can be marked in the Netherlands and Germany by a line drawn between Utrecht, Nijmegen, Düsseldorf and Hannover. And it is precisely in this North European Plain, where no rock outcrops occur, that the megalithic graves are found.[1]

The boulders used were of granite, gneiss and other rocks, and the builders simply made use of what was locally available in suitable sizes. Quartz blocks were usually absent. The larger tombs, the so-called passage graves, consist of a row of yokes or trilithons, each made of a horizontal capstone resting on two upright sidestones, the inner faces of which usually have a 10° inclination towards the interior. At each end of this row of trilithons were one or two endstones. The entrance to the elongated chamber thus formed is located in the middle of one of its long sides (Fig. 3.1). These passage graves are hence no *allées couvertes* in the French sense of the word, since the latter have an entrance at one of the narrow ends of the long chamber. Their construction is comparable, however, to that of the *sépultures à entrée latérale* of Brittany, which were contemporaneous but contained different pottery (L'Helgouac'h 1967).

TRB *dolmens*, by contrast are shorter structures, with only from one to three capstones, and they display a slightly different method of construction. In principle they are somewhat earlier than the passage graves. The term *dolmens* refers in the North exclusively to this category of tombs, and is not employed in the French sense to mean megalithic graves in general. For the latter, the terms *hunebedden* (in English also *hunebeds*) and *Grosssteingräber*, *Megalithgräber*, or *Hünengräber* are used.

The largest stones of the hunebedden are the capstones. These normally have a flat under surface and are round or elliptical in plan, and are 2-3 metres long, although larger examples occur. The three largest capstones of the impressive (but now destroyed) tomb known as König

---

[1] The absence of rock outcrops in the North European Plain lies behind the present scarcity of glacial boulders on the surface. In the 11th and 12th centuries, such boulders were used for church foundations. In the 12th and 13th century, blocks split from them (*Quadersteine*) were used for the walls of churches and castles. In the 18th and 19th century erratic boulders were used for building dikes and sluices, and in the 19th century also for macadam roads. Legal protection has preserved megalithic tombs in Drenthe since 1734. In some of the *Länder* in northwest Germany the same happened early in the 19th century, beginning with Mecklenburg (1804) and Oldenburg (1819). In former Swedish territory in Germany, the protection of ancient monuments by the Swedish law of 1666 apparently had lost its efficacy.

Fig. 3.1. Hunebeds D17 (background) and D18 at Rolde, Drenthe (postcard *c*. 1925)

Surbolds Grab or Surbolds Hus near Börger in Lower Saxony measured as much as 5-7 metres in length (Laux 1989).[2] Where capstones of unusually large or irregular form were involved, the chamber plans were adapted to accommodate them. The weight of the capstones varies from a few tons up to 20-25 tons, although the largest capstone of Surbolds Hus weighed *c*. 41-49 tons.[3] This was probably the heaviest capstone west of the Elbe. It is worth mentioning here that much larger and heavier morainic boulders existed in the North European Plain, but these apparently were too large to be manageable for hunebed construction.[4]

The supporting stones of the chambers are 1.6 to 2.7 metres tall giving internal chamber heights in Drenthe of 1.25 to 2.0 metres.[5] Internal chamber lengths at the level of the present ground surface vary here from *c*. 2.7 to 20.05 metres, and the floors are about half a metre longer.[6] The side slabs of the extremely large Surbolds Hus were *c*. 2.8 metres tall, and the interior of the chamber was *c*. 2.3 metres high and 17 metres long (Laux 1989).

The tops of several of the granite sidestones of the chamber of Langbett 686-Oldendorf 4 in the Lüneburg region had clearly been trimmed (photos in Laux 1980, figs. 11 & 13). This is also true for the sill stones, for many of the stones in the dry stone walling, and for many of the paving stones of the tombs in the TRB West Group. Fragments of a relatively rare erratic stone, a Stockholm granite, were found some 8 metres south of the southern chamber wall of hunebed D26-Drouwenerveld; other pieces of the same stone were used for the dry stone walling (analysis M.H. Huizinga).

---

[2] The internal measurements of the trapezoidal chamber of Surbolds Hus were *c*. 17 metres long and 1.6 metres to 6 metres wide. It had an east-west orientation and was surrounded by a kerb of standing slabs (Laux 1989, based on a detailed report of 1613 by Johann von Velen). There may have been a short entrance in the middle of the southern side, but another possibility, stressed by Laux, is that of a long trapezoidal chamber with the entrance at its narrow end, not unlike some of the west German gallery graves and French *allées couvertes*. Comparable tomb plans, quite exceptional for the TRB West Group, are 917-Haltern (Sprockhoff 1975) and Hilter (Schlüter 1985), both Lkr. Osnabrück (Laux 1989, figs 2-3).

[3] Using Huisman's formula for the volume: 0.6 x max. L. x max. W. x max. H. (Huisman & Van der Sanden 2003; for the measurements Laux 1989, 119). In this case 0.6 x 6.90 x 3.14 x 1.26 = 16.38m³. The weight of granite, gneiss and most other erratics varies from 2.50 to 3.00 tons per cubic metre. That of quartz sandstone varies from 2.00-2.65 tons per cubic metre (Speetzen 1998).

[4] The largest known erratic boulder in the North European Plain is the Triglaffstein at Tychowo, Pomerania Poland (760.8m³, 1978 tons). The Markgrafenstein at Rauen measured 280m³ and weighed 750 tons. The Giebichenstein at Nienburg, the largest in Lower Saxony, weighed 276 tons. The Grosser Stein at Rahden-Tonnenheide weighed originally 300

tons (Speetzen 1993; 1998). Petrus Camper (ms. 1768-1789, Amsterdam University Library) argued from the discovery of a 60 ton boulder at Helpman near Groningen that large stones were locally available for hunebed building, but that this particular one was heavier than the heaviest stone used in Drenthe hunebeds.

[5] Van Giffen 1925-1927, atlas pl. 123, 130, 137, 147: D54, D40, D30, D21, D22: height of side or end stones *c*. 1.6-2.7 metres, internal chamber heights (surface of floor stones to underside of ceiling) 1.2-1.75 metres.

[6] Bakker 1992, table 3; Van Giffen 1925-27, atlas.

Some of the flat inner sides of orthostats and capstones look as if freshly cleft, but even in oblique sunlight wedge marks have never been identified. The boulders may have been split along existing fractures by hammering. Pairs of split boulders are rare. Van Giffen (1925, p. 78) noted that capstones D5 and D6 of hunebed D27-Borger were *"apparently halves of one and the same erratic block"*. On the other hand, two half-boulders among the orthostats of hunebed D26-Drouwenerveld, which I thought were matching, proved actually to be of quite different rock types (analysis M.H. Huizinga).

Given the concentration of passage graves on the North European Plain, the erratics used to build them must have been plentifully available, but it is not possible to estimate the distances over which the stones were hauled to the sites. A sufficient number of boulders was usually available, it seems, within 350-450 metres, and distances never exceeded 5 kilometres. These erratic blocks had initially been incorporated in the impervious glacial till layers of the Saalian ground moraine, but heavy erosion during the later Eemian and the Weichselian periods had brought them to the surface, where some of them were again hidden by the deposition of Weichselian windblown coversands. It is upon these stone-free, well-drained coversands that the hunebeds were built.

That special kinds of rock, with different colours or properties, were chosen for prominent positions in the walls of the chamber, the passage or the kerb, is not obvious to me. Chamber and passage would in any case have been dark. Tall blocks were often placed as *Wächter* ('watchmen') at the corners of the rectangular kerbs around several West-German hunebeds or unchambered barrows (*Langbetten*).

A study of the small stones in and around hunebed D26-Drouwenerveld, more than 1000 in number, found that 309 of them were suitable as key rocks for an improved Hesemann count.[7] This result is almost identical to the Hesemann count of stray erratics undertaken twenty years earlier in the neighbourhood (analysis M.H. Huizinga 1970; Zandstra in Bakker in prep.). Both spectra are typical of the third and last ice stream of the Saalian glaciation in eastern Drenthe and western Westphalia, reaching from Groningen to beyond Münster (Zandstra 1993). The smaller stones were thus locally collected, as one would expect.

In the southern part of the North European Plain, near the foot of the Mittelgebirge in Germany, stones for the building of TRB megaliths were sometimes taken from identifiable rock outcrops, generally of sandstone. Sandstone blocks in TRB passage grave 981-Rheine derive from a bedrock outcrop in the Osning hills, implying transport over at least 10 kilometres (Eckert 1985). Sandstone or limestone is recorded at TRB-graves Wechte 1-2 and Recke-Espel, although the possible source outcrops have not yet been identified.[8] Gallery graves of the Wartberg Culture, which is contemporary with and related to the TRB Culture, included sandstone slabs from outcrops which were up to 0.5, 2.5, 3.5 and 4.5 kilometres away (Günther 1986, 1996; Schierhold, this vol.).

BOULDER TRANSPORT AND CONSTRUCTION METHODS

All the heavy boulders were probably transported to the building site on sledges (forked tree stems or rectangular sleds) that were drawn with ropes and propelled with the help of levers, perhaps over rollers. Loose ground surfaces may have been strengthened by transversely laid branches if they were not frozen. The sledges were pulled by large numbers of people. Ox traction is a less probable option; on Sumba, an island near Bali to the east of Java where the building of megaliths remains a living reality, buffaloes are used for sacrifices to the gods and to feed the hauliers, but not for traction.

Roger Joussaume (1985) and others have compiled depictions on stone walls in ancient Egypt and Assyria and drawings and photographs on paper of such stone dragging. Estimates of the minimum number of people needed to transport the blocks vary from one person per ton on a smooth, level and lubricated clay surface to ten people per ton on a natural 9° slope (Arnold 1991, 63).

On Sumba, Dr Gerard J. Onvlee took a great many photographs documenting the transport of a massive stone cist and its capstone at Waikabubak in about 1968. It is worth remembering, as archaeologists, that here it is not the minimum number of people needed to move the stones that is important, but the maximum number of people that can take part and be fed and entertained, to ensure that the dead person's and the builder's name "will live for ever".

At the building site, the side stones of the hunebed chamber were tilted and lowered over the edge of the foundation socket into a packing of cobbles and sand. The side stones had a flat inner face and their form was more or less that of an obliquely halved hardboiled egg. Their centre of gravity is towards the bottom of the outer face and this made it possible for them to stand inclined at an angle of 10° towards the interior of the chamber. The earth on the inner side of the foundation socket may have been reinforced by standing stakes to take the pressure of the stone as it was tilted into position.[9] Following this, the

---

[7] A Hesemann count aims at identifying the origin of the ice stream by which the erratics were deposited. The numbers and sorts of local key erratics compiled in a Hesemann formula reflect one of the regions in Fenno-Scandia from where they came. An improved formula is given by J.G. Zandstra (1992). Regions with similar Hesemann spectra derived their erratics by the same ice stream.

[8] But large sandstone boulders also occur as erratics in Westphalia (Speetzen 1998).

[9] The discoloration under the floor of hunebed D26-Drouwenerveld probably corresponded to one of these (Bakker, in prep.).

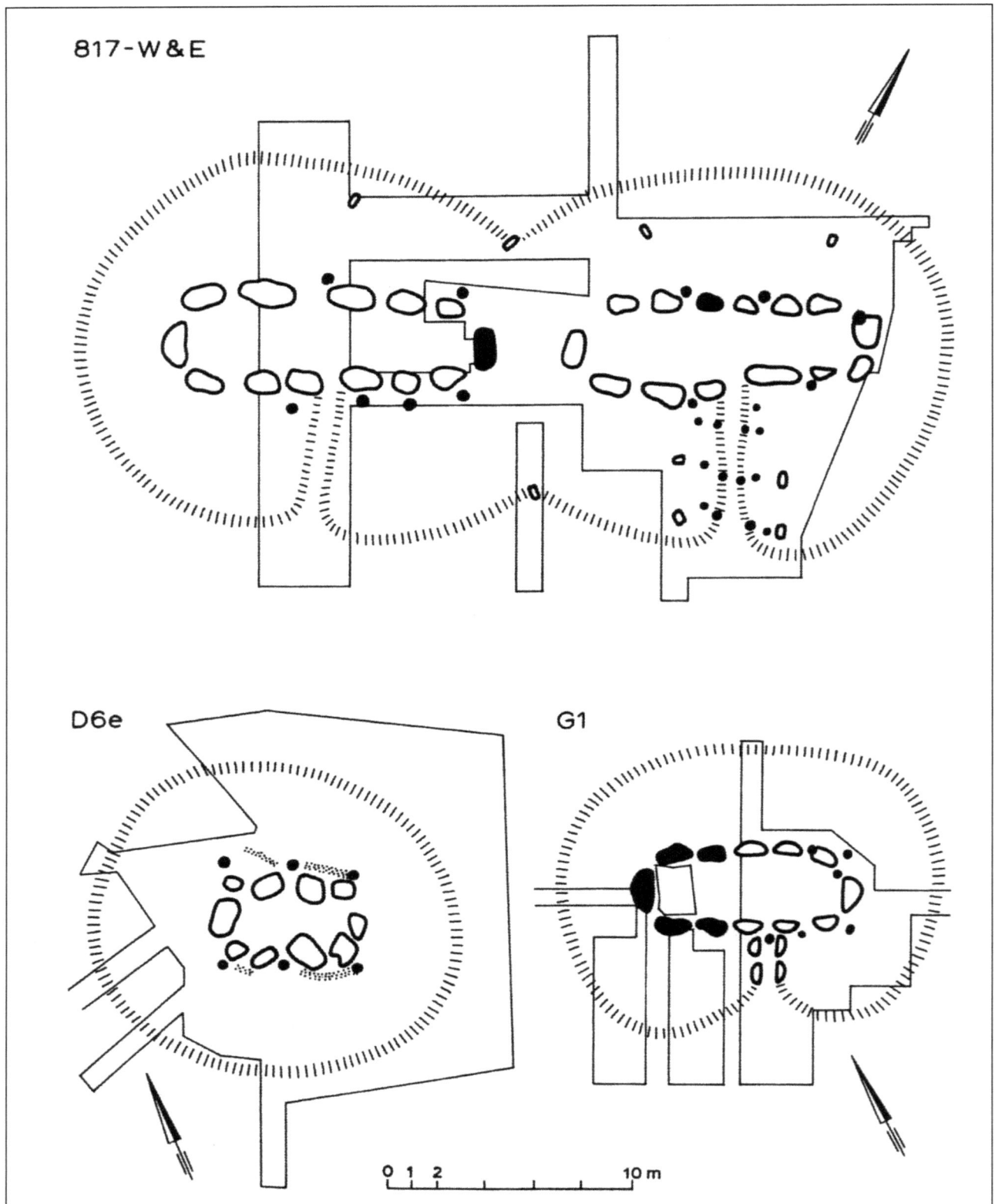

Fig. 3.2. Plans of hunebeds at Tannenhausen, northwest Germany (817-W & E) (above) and D6e-Tinaarlo and G1-Noordlaren in the Netherlands, showing postholes, possibly traces of wooden ramps for positioning the capstones (Bakker 1992, fig.17; 1999, fig.4)

floor of the chamber was excavated to the level of the bases of the inner sides of the side stones, and the floor was then covered with a paving of cobbles and broken stones (up to c. 45 cm across).

The gaps between the uprights were filled by dry-stone walling. Dehn, Hansen and Westphal have found excellent examples of dry stone walling in Denmark made of high quality flagstone or layered limestone. In the Netherlands and northwest Germany, however, the builders had to make do with broken stones and flattish cobbles up to 45 cm in diameter. The observation of Dehn and Hansen (2006) that the dry stone walling was completed before the lower part of the mound was finished and the capstones placed in position may also apply to these tombs west of the Elbe.

The photographs taken by Onvlee show that the capstones of the Sumba tombs were placed in position with the help of a lightly-built wooden ramp.[10] This is confirmed by other photographs and detailed descriptions of tomb construction made on Sumba since 1913. Other photographs depict the dragging of a 4m³ menhir by 525 men up a natural 30-40° slope at Buwomataluo on Nias, west of Sumatra, in 1914 (ills. in Schröder 1917, Röder 1948; cf. Bakker 1992; 1999, figs 6-8). I agree with Röder (1948) that the Sumba method with wooden ramps may very well have been used for the megalithic tombs of the TRB West Group.[11]

There is, however, a complication. At a few of the hunebedden in Drenthe and German Ostfriesland, substantial post holes have been noted around the chamber, under the base of the mound. I have interpreted them as traces of the scaffolding of a wooden ramp (Fig. 3.2). Why wooden posts of the TRB culture are sometimes visible as soil discolorations, but are generally invisible in the non-calcareous sands of the region, remains unclear. The post holes that survive do appear, however, to have held the extremities of substantial and fairly permanent posts fixed in the sand. On Sumba, the wooden scaffolds are instead rather flimsy bamboo constructions, which stood *upon* the ground surface and were immediately removed after the placing of the capstone. How did these timber posts operate, then, in Drenthe? Were the apparently substantial wooden props retained within the mound after the construction of the tomb, and if so for what purpose?

THE MOUND

Sections through a hunebed mound drawn in 1809 (Fig. 3.3) show that the gaps between the capstones were filled from the outside with smaller stones and covered by at least 15-35 cms of sand. These sections of hunebed Emmen-41 are the only complete sections through an intact hunebed mound ever drawn in Drenthe. One would expect the stones between and adjacent to the capstones to have been covered with a layer of loam and the dry-stone walling of the chamber walls to have been packed in loam, so as to divert rainwater sideways and keep the chamber completely dry. This at least is the practice that has been documented in well-preserved Danish tombs (Dehn & Hansen 2006), but it was not reported in 1809 by the keen local officials and the government draughts-man.[12] Furthermore, nowhere has loam been noticed in the mound sand of excavated tombs, although it was plentifully available in the region, and was used for instance for the sublime TRB ceramics.

Fergusson (1872, 321) signalled another major problem in seeking to understand the original form of the mounds of the larger Drenthe hunebeds:

"The first question that arises with regard to these Hunebeds is, were they originally covered with earth or not? That some of the smaller ones were and are is clear enough, and some of medium size are still partially so; but the largest, and many of the smaller, do not show a vestige of any such covering; and it seems impossible to believe that on a tract of wretched barren heath, where the fee-simple of the land is not now worth ten shillings an acre, any one could, at any time, have taken the trouble to dig down and cart away such enormous mounds as would have been required to cover the monuments. It seems here clearer than almost anywhere else, that, even if it had been intended to cover them, that intention, in more than half the cases, was never carried into effect."

The ruinous state of the other Drenthe passage graves hampers a discussion of their original mound construction. This especially so since whatever remained of these mounds was removed during so-called restorations in the early 1870s, when tens of centimetres of mound material were taken away, resulting in some cases in their complete removal (D15-Loon), without any scientific supervision (Fig. 3.4-3.5; Bakker 1979).

Lanting (in press) concludes that in several Dutch hunebeds the mound reached only slightly higher than the tops of the uprights. Where the podsol soil profile – formed roughly between 2000 BC and the Roman period – is sufficiently well preserved for the form of the mound to be reconstructed, its slope never reaches much higher than that. The mounds may also have covered the stones inserted in the gaps between the capstones. But they did not cover the tops of the capstones,[13] unlike the hunebed

---

[10] Photographs by H. Witkamp in eastern Sumba in 1913 and by G.J. Onvlee in Waikabubak in western Sumba in 1968 show the general use of wooden ramps. A recent photo from Kodi, West-Sumba (R. Adams, this vol.), however, shows the raising of a capstone using levers on a growing pile of inserted beams in a densely crowded cemetery, probably because it took less space and labour.

[11] In addition to the studies on transport and construction of megalithic graves on Sumba, and the transport of a large menhir on Nias (to the west of Sumatra) in 1914 (summarised in Bakker 1992 & Bakker 1999), Adams 2004 and Adams (this vol.) describes the sociological back-grounds of these procedures in western Sumba. See also the extensive series of photographs by G.J. Onvlee showing the transport and positioning of capstones in Waikabubak, *c.*1968.

[12] The number of stones in the crevices between the capstones is rather small in the 1809 drawing, neither were the cobbles above the capstones flat. Perhaps this was due to destruction by the finders, stone-seekers, before the officials arrived.

[13] The presence of podsol layers in this position excludes the possibility that the upper parts of mounds as tall as that shown in the 1809 sections had been removed by stone seekers in the Middle Ages or later.

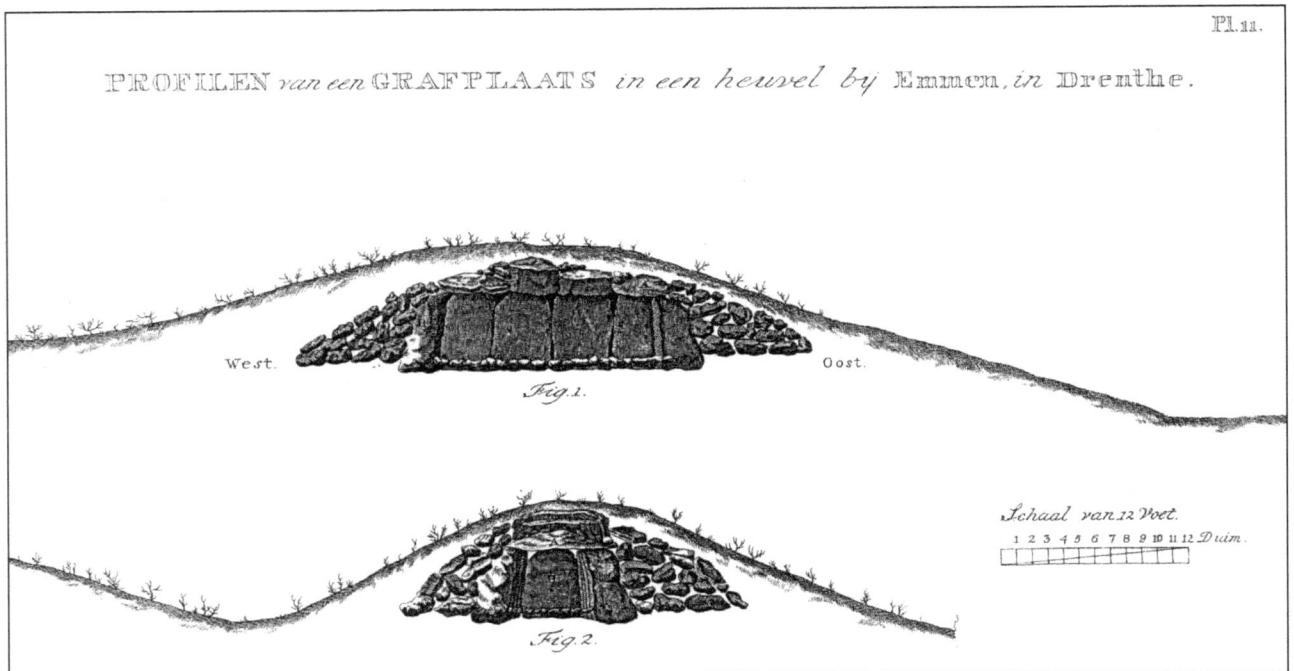

Fig. 3.3. Hunebed D41 at Emmen, Drenthe, as discovered under an earthen barrow in 1809. Print carefully reproduced from the 1809 drawing by the surveyor P.A.C. Buwama Aardenburg (in Westendorp 1815)

Fig. 3.4. Hunebed D15 at Loon, Drenthe, in 1847 (after drawing by L.J.F. Janssen, Drents Museum, Assen)

excavated in 1809 at Emmen and the superb Danish tombs reported by Dehn. In 1847, several Drenthe hunebeds were still "*surrounded by earth as high as the capstones, so that these tombs look as if they were just a group of flat boulders lying in a row on a hillock*" wrote Janssen (1853). Janssen's drawings of 1847 (e.g. Fig. 3.4) and Fergusson's remarks of 1872 confirm this.[14]

To this may be added the 'grand dolmen' G5-Heveskes-/klooster (northeastern Netherlands) and several passage graves of medium size at Wanna near Cuxhaven in Germany that were discovered in the 20th century under a prehistoric peat layer. They had only very low mounds that did not even reach the top of the uprights. The mound material consisted only of what had been produced by the digging of the foundation pits and the chamber floor.

This leaves us with the still unanswerable question: why did so many of the West Group chambers have no

---

[14] The 1809 tomb may have been completely covered by its mound because the capstones were thin and flat, whereas other tombs generally have bulky boulders as capstones.

Fig. 3.5. Hunebed D15 at Loon, Drenthe, in 1874 (photo J. Goedeljee, Pleyte archives, State Museum of Antiquities, Leiden). The mound was removed in 1870, as the lichen-free parts of the stones show; even the packing stones of some uprights were exposed (foreground)

protective layer of loam and stones, or at least a much less stable one than the Danish tombs? Could it be that the Western TRB builders did not even aim at a perfectly dry chamber interior? That is hard to believe, although it is the case that they buried the greater part of their dead in earthen flat graves.[15]

I have sought here to derive general principles from the scanty data available on hunebed construction, as is the usual practice in science. But in several respects the Dutch data differ somewhat from the observations in Denmark, reminding us of the fact that similar things need not always have been brought about in exactly similar ways, as several colleagues have remarked. This may be especially true for the TRB culture which in many other aspects excelled in an amazing variety.[16]

References

ADAMS, R. (2004): – *The megalithic transport of West Sumba, a preliminary report*. www.sfu.ca/archaeology/dept/fac-bio/hayden/index. htm.

[15] Most of the older literature and related data is referenced in Bakker 1992 & 1999, to which the current paper gives later additions.

[16] I am most grateful to Dr. Gerard J. Onvlee, Leusden, for organising his slide collection from Waikabubak, West Sumba (*c.* 1968), placing it on the internet www.onvlee.nl/wims/php/slideshow.php, and for additional information and his permission to publish some photographs here. Dr. Onvlee was for many years medical superintendent of the Waikabubak hospital.

ARNOLD, D. (1991): – *Building in Egypt, Pharaonic stone masonry*, New York-Oxford: Oxford University Press.

BAKKER, J.A. (1979): – 1878, Lukis and Dryden in Drente. *The Antiquaries Journal* 59, 9-18.

BAKKER, J.A. (1992): – *The Dutch hunebedden, megalithic tombs of the Funnel Beaker Culture*, Ann Arbor: International Monographs in Prehistory.

BAKKER, J.A. (1999): – The Dutch megalithic tombs, with a glance at those of North-West Germany. In Beinhauer, K.W., Cooney, G., Guksch, G.E., & Kus, S. (eds) *Studien zur Megalithik. Forschungsstand und ethnoarchäologische Perspektiven / The Megalithic phenomenon*, Mannheim-Weissbach: Beier & Beran, 145-162.

BAKKER, J.A. (in preparition): – *Hunebed D26 in het Drouwenerveld. Verslag van de onderzoekingen*.

DEHN, T.; HANSEN, S.I. (2006): – Birch bark in Danish passage graves. *Journal of Danish Archaeology* 14, 23-44.

ECKERT, J. (1985): – Rheine. *Ausgrabungen und Funde in Westfalen-Lippe* 3, 402-403.

FERGUSSON, J. (1872): – *Rude Stone Monuments of all Countries. Their Age and Uses*. London: John Murray.

GIFFEN, A.E. VAN (1925-1927): – *De hunebedden in Nederland*. Utrecht: A. Oosthoek.

GÜNTHER, K. (1986): – Ein Grosssteingrab in der Warburger Börde bei Hohenwepel, Stadt Warburg, Kreis Höxter. *Ausgrabungen und Funde in Westfalen-Lippe* 4, 65-144.

GÜNTHER, K. (1997): – Die Kollektivgräber-Nekropole Warburg I-V. *Bodenaltertümer Westfalens* 34, 163ff.

HUISMAN, H., & SANDEN, W.A.B. VAN DER (2003): – 'Kolossale steenen' onder het veen. Het vermeende hunebed van Nieuw-Dordrecht'. *Nieuwe Drentse Volksalmanak* 120, 129-137.

JANSSEN, L.J.F. (1853): – Over de beschaving der allervroegste bewoners van ons vaderland (eene archaeologische voorlezing). *Oudheidkundige Verhandelingen en Mededeelingen van Dr. L.J.F. Janssen* I. Arnhem, 1-26.

JOUSSAUME, R. (1985): – *Des dolmens pour les morts, les mégalithes á travers le monde*. Paris.

LANTING, J.N. (in press): – Some observations of the construction of Dutch megalithic tombs, *Proceedings of the conference on megalithic tombs at Kalundborg, 1995*.

LAUX, F. (1980): – Die Steingräber auf der Lüneburger Heide. In Körner, G. & Laux, F. (eds) *Ein Königreich an der Luhe*. Lüneburg: Museumsverein für das Fürstentum Lüneburg, 89-218.

LAUX, F. (1989): – König Surbolds Grab bei Börger im Hümmling, *Nachrichten aus Niedersachsens Urgeschichte*. 58, 117-127.

L'HELGOUAC'H, J. (1967): – Les sépultures à entrée latèrale en Armorique. *Palaeohistoria* 12, 259-281.

RÖDER, J. (1948): – Bilder zur Megalithentransport. *Paideuma* 3, 84-87.

SCHLÜTER, W. (1985): – Das Grosssteingrab von Hilter am T.W., Lkr. Osnabrück. In Wilhelmi, K. (ed.) *Ausgrabungen in Niedersachsen. Archäologische Denkmalpflege 1979-1984*. Beiheft I. Stuttgart, 122-130.

SCHRÖDER, E.E.W.G. (1917): – *Nias. Ethnografische, geografische en historische aanteekeningen*, Leiden: Brill.

SPEETZEN, E. (1993): – Grossgeschiebe (Findlinge) in der Westfälischen Bucht und angrenzenden Gebieten und ihre Bedeutung für die Eisbewegung. In Skupin *et al.* (eds), *Die Eiszeit in Nordwestdeutschland. Zur Vereisungsgeschichte der Westfälischen Bucht und angrenzender gebiete*. Krefeld: Geologisches Landes-/amt Westfalen, 34-42.

SPEETZEN, E. (1998): – *Findlinge in Nordrhein-Westfalen und angrenzenden Gebieten*. Krefeld: Geologisches Landesamt Westfalen.

SPROCKHOFF, E. (1975): – *Atlas der Megalithgräber Deutschlands III, Niedersachsen-Westfalen*. Bonn: Rudolf Habelt.

WESTENDORP, N. (1815): – Verhandeling ter beantwoording der vrage: Welke volkeren hebben de zoogenoemde hunebedden gesticht. In welke tijden kan men onderstellen, dat zij deze oorden hebben bewoond? *Letter- en Oudheidkundige Verhandelingen van de Hollandsche Maatschappij der Wetenschappen te Haarlem* I, 233-377.

ZANDSTRA, J.G. (1993): – Nördliche kristalline Leitgeschiebe und Kiese in der Westfälischen Bucht und angrenzenden Gebieten. In Skupin *et al.* (eds), *Die Eiszeit in Nordwestdeutschland. Zur Vereisungsgeschichte der Westfälischen Bucht und angrenzender gebiete*. Krefeld: Geologisches Landesamt Westfalen, 43-143.

THE GALLERY GRAVES OF HESSE AND WESTPHALIA, GERMANY: EXTRACTING AND WORKING THE STONES

Kerstin SCHIERHOLD

Römisch-Germanische Kommission des Deutschen Archäologischen Instituts,
Palmengartenstrasse 10-12, 60325 Frankfurt am Main, Germany

Abstract: Recent research on the gallery graves of Hesse and Westphalia has for the first time included geological analysis of the stones, focusing on five surviving tombs in the valley of the river Altenau, near Paderborn, and the well known tomb at Züschen. The results allow the identification of the material and its provenance, and reveal the distance over which it was transported. Few examples offer better evidence for the quarrying of the stones. The material itself demonstrates the craftsmanship and geological knowledge of the builders who constructed these long-lasting places for the dead. These issues will be illustrated by a discussion of the Paderborn tombs and other examples.
Key-words: Hesse, Westphalia, gallery graves, geological determination

Résumé: Pour cinq allées couvertes dans le vallée d'Altenau près de Paderborn en Westphalie et pour la tombe connue de Züschen en Hesse, des études géologiques pouvaient être faites pour la première fois. Ceux-ci ont donné explications sur la provenance, la distance de transport et la disposition du matériau de construction. Des déclarations plus précises concernant la constitution des lieux d'exploitation pouvaient être adoptées dans quelques cas. Le matériau de construction lui-même permet une vue dans la capacité artisanale et la connaissance des architectes sur le secteur de la construction de tombe. Celui-ci doit être illustré aux tombes de Paderborn ainsi que d'autres exemples.
Mots-clés: Hesse, Westphalie, allées couvertes, études géologiques

This article focuses on certain aspects of the gallery graves of Hesse and Westphalia in Germany (Fig. 4.1), paying special attention to the petrology of the stones. The geological analyses on which it is based were undertaken by geologists Dr Martin Hiß and Dr Jochen Farrenschon from the Geological Service of North-Rhine-Westphalia, to whose expertise I am indebted. These investigations provide information about the building materials and allow conclusions to be reached about the quarry sites and the routes of transport. Furthermore, it is possible to ascertain not only the properties of the building materials, but also those of the quarry sites. Here some of the results will be presented and placed in the context of the group as a whole.[1]

The Hessian and Westphalian tombs are located in the area of the Funnel Beaker and Wartberg cultures (Schrickel 1966; Günther 1997; Raetzel-Fabian 2000). They were built between 3400 and 3000 cal BC and used until 2800, or sometimes 2700 cal BC. Approximately 40 gallery graves and related forms in Hesse, Westphalia and neighbouring areas are known today. The tombs measure 10 to 35 metres long and 2 to 3 metres wide, are sunk into the ground, and were covered by a mound. Access to the chamber is via a porthole entrance. The building material in the majority of cases consists of rectangular slabs of limestone or sandstone, although basalt, erratic blocks, and tertiary quartzite were also used. The tombs can be divided into two main types (Günther 1987, 92). The so-called "Züschen" type is characterised by an axial entrance in the narrow end-wall preceded by an antechamber or vestibule. The so-called "Rimbeck" type has an entrance more or less

Fig. 4.1. Map of Germany showing the areas of investigation. (Map: *The World Factbook*)

in the middle of one of the longer sides, giving a T-shaped plan similar to that of Funnel Beaker passage graves.

HESSE: "LOHNE-ENGELSHECKE I" OR ZÜSCHEN I, SCHWALM-EDER KREIS

The famous tomb of Züschen I (Boehlau & Gilsa 1898; Kappel 1989) is situated near Kassel, in the Westhessian

---

[1] The geological analyses are part of ongoing doctoral research on the gallery graves of Hesse and Westphalia. The full results will be presented in the completed thesis.

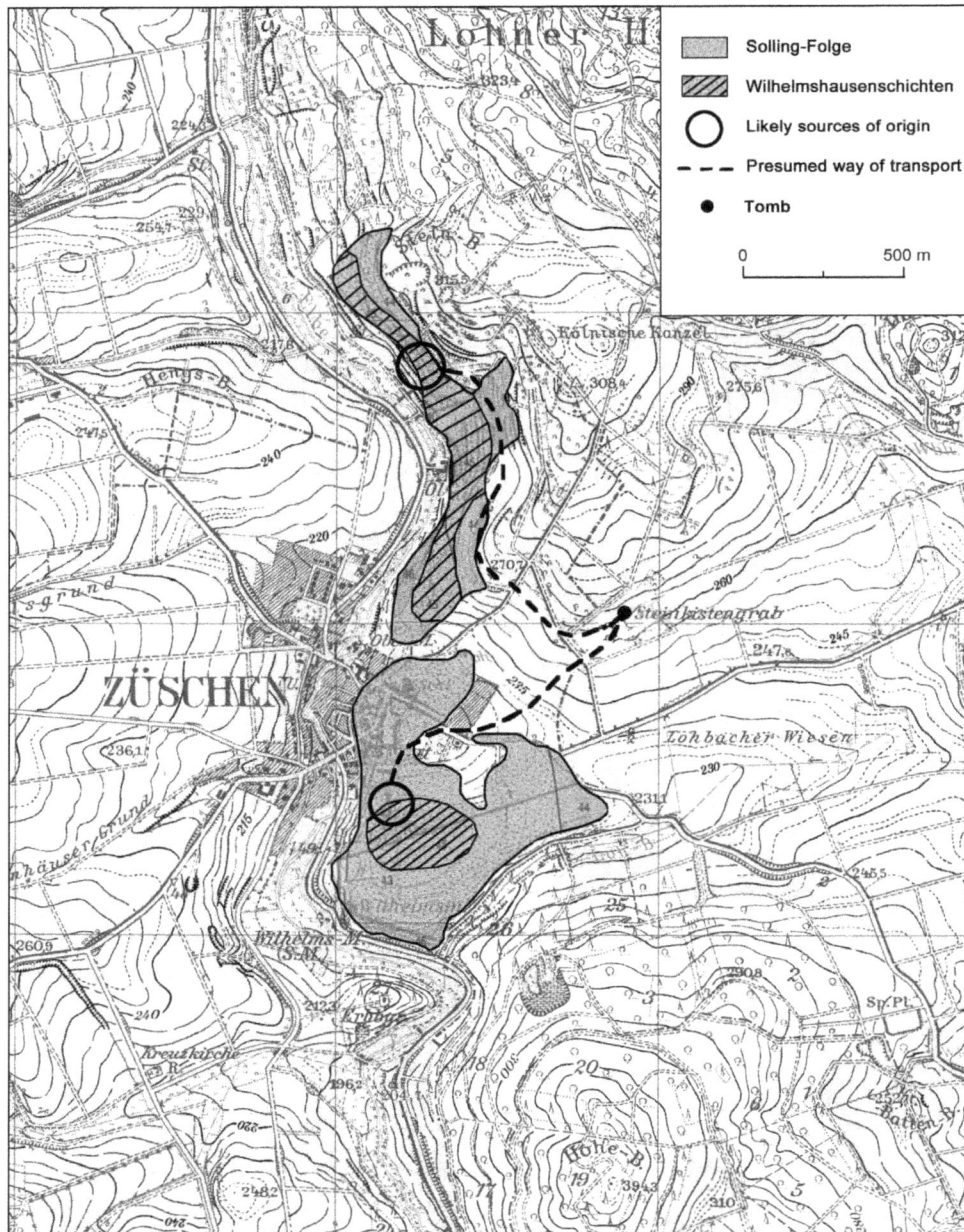

Fig. 4.2. The Züschen tomb and surroundings showing geology and presumed routes of transport.
(Topography: © German Federal Office for Cartography and Geodesy 2004)

Lowland, near a small river which subsequently joins the River Eder (Fig. 4.2). The Westhessian Lowland is characterised by Triassic surfaces which are interrupted by ridges of basalt. These now tree-covered hilltops are a prominent feature of the landscape. In the Neolithic period, they were settled by people of the later Wartberg culture (Schwellnus 1979; Raetzel-Fabian 2000).

The Züschen tomb, the eponymous example of its type, is 20 metres long and has a width of approximately 3.5 metres (Fig. 4.3A). It is subdivided into an antechamber (2.5m long) and chamber (16.5m). Twenty-five stone slabs are preserved, and these were geologically examined and analysed. The composition of the slabs shows that they are from the so-called Middle Triassic or "Mittlerer Buntsandstein". Closer inspection of the stones allows them to be attributed to the "Wilhelmshausen-Schichten" of the so-called "Solling-Folge" within this formation. The stones are yellowish to brownish-grey in colour; their texture is coarse and sandy.

Six outcrops of the "Wilhelmshausen-Schichten" are visible close to the tomb. Two areas (judging from their proximity to the tomb and the special properties of the beds, which match the thickness of the stones used) seem most likely to have been the loci of Neolithic quarrying. They are situated north and south of the tomb near the scarps of the small river. In retracing the distance and the

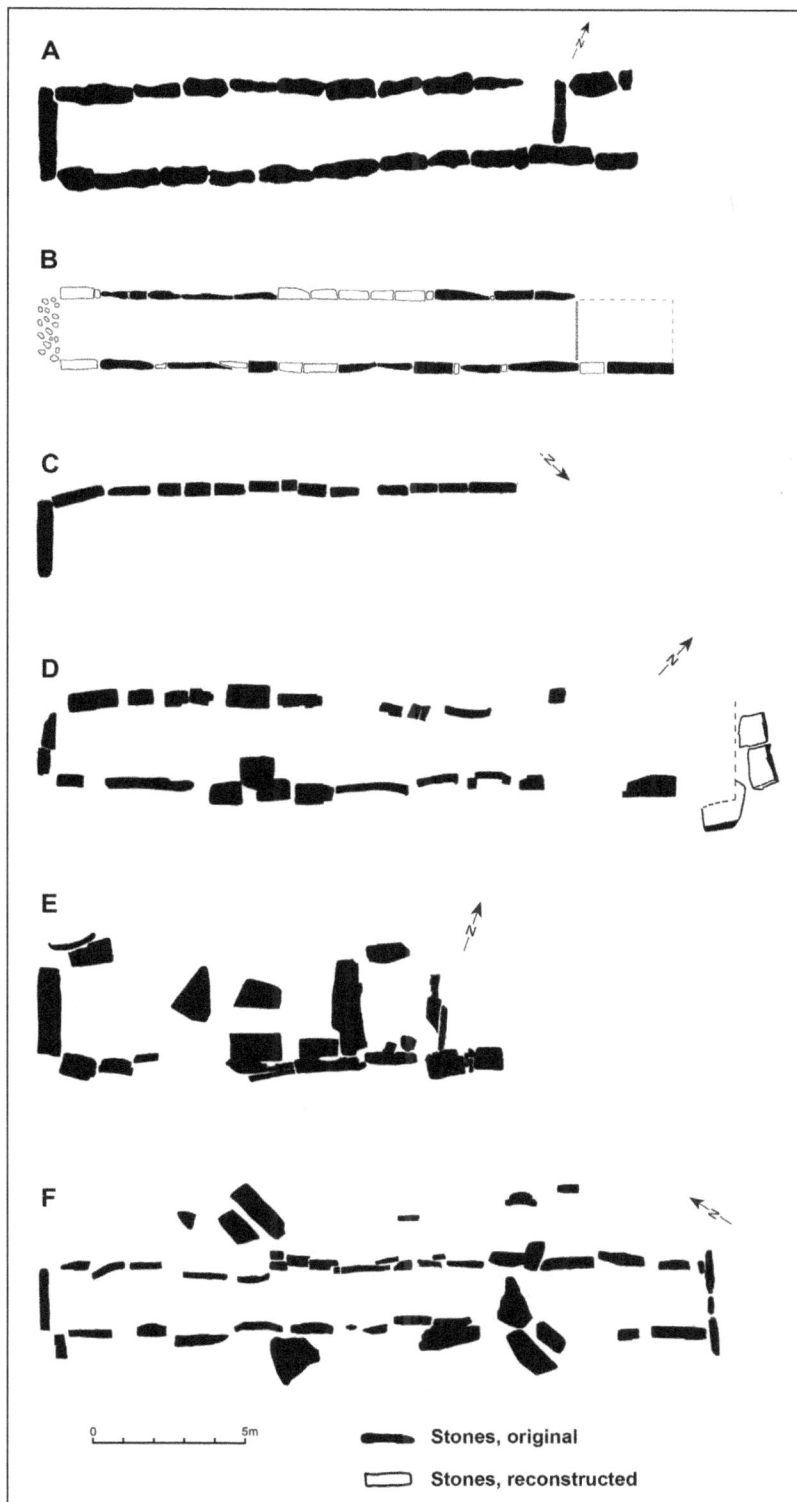

Fig. 4.3. Plans of the tombs discussed. A: Züschen; B: Atteln I;
C: Henglarn II; D: Kirchborchen I; E: Kirchborchen II; F: Etteln

route of transport, it is assumed that the route with the smallest difference in altitude was chosen, even where this was longer than the direct route (Günther 1987, 83-84; Günther 1997, 180). From the northern area to the tomb the route of transport from outcrop to monument measures *c*.1300 metres. The southern route, covering a distance of 1000 metres, is shorter, but the northern route is less steep than this southern route and is therefore more likely to have been used since it offered an easier method of access (Fig. 4.2).

Fig. 4.4. Outcrop of the Wilhelmshausen-Schichten south of the Züschen tomb (Photo: Dr J. Farrenschon)

PROPERTIES OF BUILDING MATERIALS AND QUARRYING SITES

The "Wilhelmshausen-Schichten" are exceptional good for providing slab-like building materials. The beds are naturally separated by horizontal and vertical fractures which give varied thicknesses. Owing to their sandy texture, blocks could extracted from the beds with relative ease by the use of simple tools such as wooden wedges and chisels. The fractures predetermined the form of the slabs. In the Neolithic period it can be safely assumed that it was the uppermost beds, closest to the surface that were exploited. These bear deep marks and cracks from long-term weathering that would have facilitated the quarrying of the slabs (Fig. 4.4).

At Züschen the majority of the slabs have the same thickness (about 0.5 metres), and it can therefore be assumed that they were extracted from the same bed of sandstone. Two different slabs with thicknesses of approximately 0.2 m and 0.8 metres respectively are assumed to come from other beds. The softness of the stones allows the surfaces easily to be shaped or smoothed. It would not have been too difficult therefore to cut the 50 cm diameter porthole entrance through one of the slabs, probably with a stone tool, and to carve motifs on the slabs.[2]

WESTPHALIA: THE ALTENAU VALLEY, KREIS PADERBORN

A group of gallery graves is located in the Altenau valley close to the city of Paderborn, in North Rhine-Westphalia (Fig. 4.5). The valley of the Altenau falls within the Paderborn plateau, which has an Upper Cretaceous geology and is characterised by dry plateaux, deep incised arroyos and karst springs. Seven megalithic tombs are known today in the Altenau valley, and geological examination was undertaken at five of these sites.

Atteln I

The only reconstructed tomb is Atteln I, the southernmost tomb, situated on the Altenau valley floor (Günther 1979) (Fig. 4.5). The tomb measures 21 metres long and has a width of 2.5-3 metres, with an internal width of 2 metres, and an internal length of 19.5 metres. It belongs to the Züschen type of gallery grave with an axial entrance, although there is no antechamber (Fig. 4.3B). Sixteen of the original 17 stones have been studied, but one of them could not be analysed owing to its deeply embedded position at the western side of the reconstructed monument. Through fossil inclusions the stones can be assigned to the Cenomanian (first stage of the Upper

---

[2] See A. Jockenhövel & L. Loerper 'Iconography and optical 3D measurements techniques: a modern view on the megalithic art of the

gallery grave at Züschen/Lohne (Germany)' (paper presented in session C 26: *Prehistoric Art: Signs, symbols, myth, ideology* at UISPP Conference, Lisbon 2006).

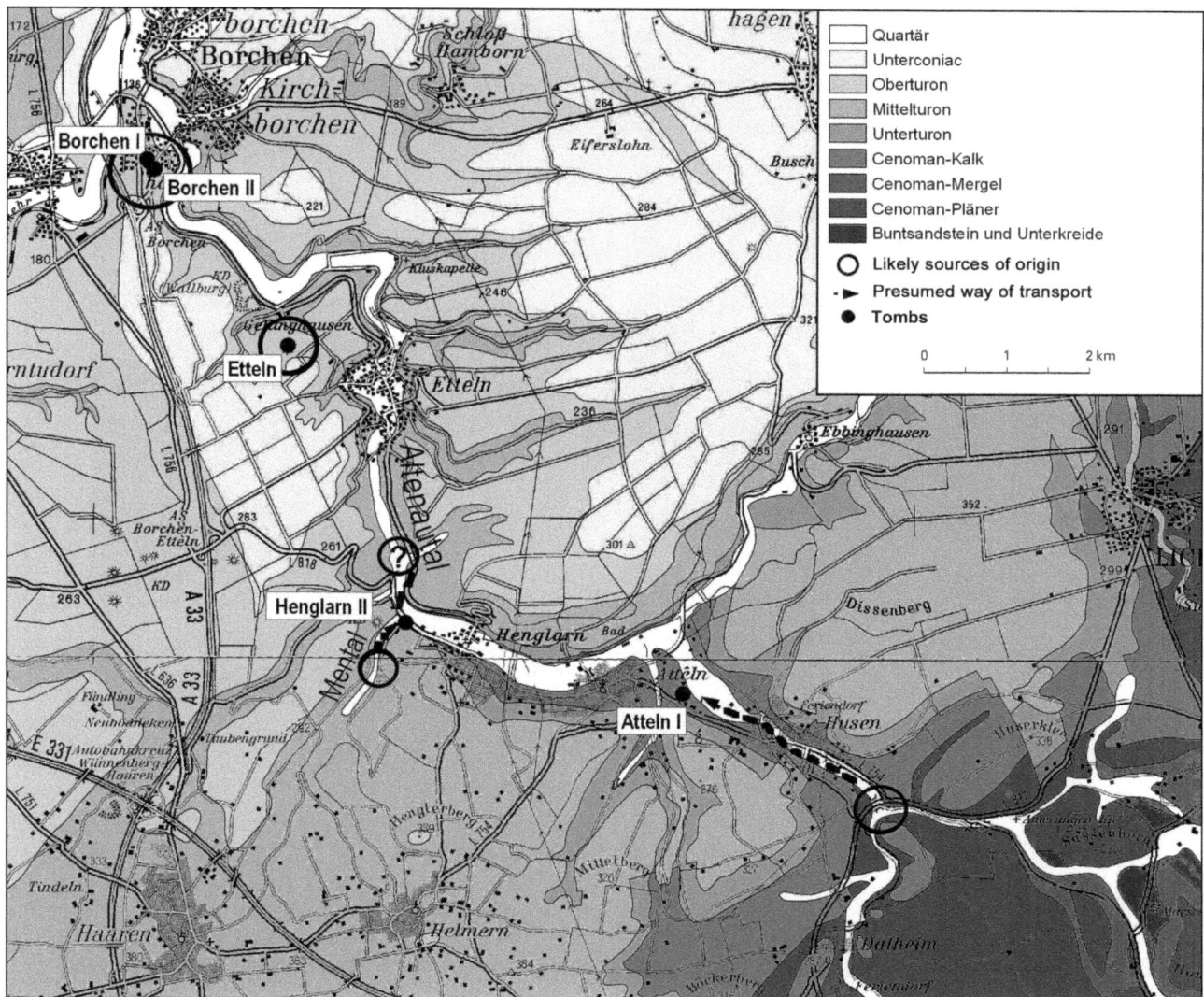

Fig. 4.5. The Altenau valley showing geology and presumed routes of transport. (Geology: Vervielfältigt mit Genehmigung des Geologischen Dienstes NRW - Landesbetrieb - vom 24.10.2006; Geologische Karte von Nordrhein-Westfalen 1: 100.000 mit Erl. - Hrsg. Geol.-L.-Amt Nordrh.-Westf.; Krefeld. Blatt C 4318 Paderborn, 2. Aufl. (1985), Bearb. DAHM, H.-D. *et al*. Topography: © German Federal Office for Cartography and Geodesy 2004)

Cretaceous). More precisely the beds belonged to the so-called "Pläner" limestone and the so-called "Flaserkalk-stein-Folge". Their colour is grey, but many stones show a yellow to brown coloured surface as a result of weathering. Their texture is slightly corrugated and smooth, but crumbly and irregular at the edges.

In seeking the outcrops from which these blocks derive, we may note that beds of the "Pläner" and "Flaserkalkstein-Folge" of the Cenomanian are visible at the confluence of the Altenau and Piepenbach rivers, immediately southeast of the tomb. The most accessible path with the least difference in altitude from source to monument is provided by the riverbed of the Altenau. The distance from outcrop to tomb measures some 2700 to 2800 metres. Today, the confluence of the rivers Altenau and Piepenbach is masked by a flood control retention basin, but a few areas of the former land surface are visible and accessible. Suitable slabs similar to those used for building the tomb can be found on the scarp.

Henglarn II

This tomb lies at the border of the floodplain where the arroyo of the River Men meets the valley of the Altenau (Günther 1980) (Fig. 4.5). Only one of its longer sides has survived, the other half having been truncated and destroyed during road construction in the 19th century. An erratic block lies nearby and may have been part of the tomb structure. The tomb probably belongs to the Rimbeck type of gallery grave, but owing to the destruction it has suffered this is difficult to confirm. The tomb is 18.5 metres long and, judging from the sole preserved stone on the eastern side, approximately 2.5 metres wide (Fig. 4.3C).

Fig. 4.6. Bed of the River Men: a possible quarry site (Photo: K. Schierhold)

The fourteen stones examined consist of Middle Turonian limestone from the so-called "Lamarcki-Schichten". The Turonian forms the second stage of the Upper Cretaceous. These stones have extremely smooth surfaces especially along the natural fissures. Middle Turonian limestone is available very close to the tomb, and the slabs were probably taken from the Altenau or Men valley. The routes of transport along the valley floor or the riverbed would have been very short, from 50 metres to a maximum of 600 metres. In the Altenau riverbed the limestone is overlain by medieval alluvial deposits, and it is not possible to identify any Neolithic quarry site. In the valley of the river Men, however, the quarry sites are still accessible. Slabs of limestone, identical in size to those used in the tomb, can be found both on the banks and in the riverbed itself (Fig. 4.6).

Kirchborchen I and II

The tombs of and Kirchborchen I and II (Günther 1976; 1978) are presented here together, because they are made of the same building material and lie only 120 metres apart (Fig. 4.5). Unlike Atteln I and Henglarn II they are situated not on the flood plain but on the northern slope of the Limberg, a hill on the western edge of the Altenau valley. Kirchborchen I belongs to the Rimbeck type with an entrance in one of its longer sides. It is approximately 23 metres long and 3.5 metres wide, the chamber itself measuring some 22 metres long and 2.5 metres across (Fig. 4.3D). Twenty-one of the stone slabs were geologically examined. The tomb of Kirchborchen II is of

Züschen type with axial entrance and antechamber. It is 13.5 metres long and 3.8 metres wide, measuring approximately 11.5 metres long internally with a 2.8 metre antechamber (Fig. 4.3E). In this case, 22 of the slabs were analysed. The survival of several of the capstones is a notable feature of both tombs.

All the stones used in these tombs are limestone slabs from Upper Turonian formations, in particular from the so called "Striatoconcentricus-Schichten". Characteristic of these slabs are chert inclusions, which are seldom found in the Upper Turonian, and have not yet been located in the Kirchborchen area. The two thin porthole-slabs of Kirchborchen II come from a particular bed within the Upper Turonian, the so called "Soester Grünsand", which contains abundant quantities of the green mineral glauconite.

Beds of Upper Turonian limestone may be found in the immediate vicinity of the tombs and very close to the surface. A source for the thin slabs of Soester Grünsand was located some 250 metres from Kirchborchen II. It seems very likely therefore that most of the stones were quarried directly at the site of the tombs.

Etteln

The Etteln tomb (Günther 1978) is located in an area similar to that of the Kirchborchen tombs. It stands on the Lechtenberg on the western edge of the Altenau valley (Fig. 4.5). The tomb is 22 metres long and 2 to 2.5 metres

Fig. 4.7. Karst groove on a slab from Atteln I (Photo: Dr M. Hiß)

wide; its internal measurements are 21.5 x 2.3-2.8 metres. It belongs to the Züschen type of gallery grave, although there is no axial antechamber (Fig. 4.3F). Forty-five slabs, among them several capstones, were closely examined. As at Kirchborchen, the composition of the stones shows that they came from the "Striatoconcentricus-Schichten" of the Upper Turonian. Some of them have the same chert inclusions. Upper Turonian beds occur in immediate proximity to the tomb and very close to the surface, and here again therefore it seems very likely that the slabs were extracted close to the site where the tomb was built.

The materials and quarry sites

The building material used for all the tombs in the Altenau valley is limestone from two different stages of the Upper Cretaceous, both with similar properties. The bedding planes and fracture lines divide them into rectangular and trapezoidal slabs and blocks that could easily be extracted using wooden wedges and chisels. As at Züschen, it is quite certain that only the uppermost beds were exploited. All of the stones that were examined show natural and unworked surfaces, with edges marked by natural fissures. Only the porthole-slabs have been shaped. In the case of Kirchborchen I and II and Etteln, the lower corners of the door-slabs are cut away, so that the porthole looks roughly like an arch.

The stone slabs used in the monuments give further clues to the specific properties of the quarry sites. Slabs of varied thickness were utilised at Atteln I and Kirchborchen I and II. This indicates that more than one

bed of limestone was quarried, probably in different areas, although it is possible that beds of varying thickness overlay one other. On the other hand the chert inclusions and the almost identical thickness of several slabs used at both Kirchborchen I and Kirchborchen II point to their extraction from a single specific bed of limestone. Upper Turonian limestone is present in the immediate vicinity of the tombs and can be found close beneath the surface. Hence it seems very likely that these slabs were quarried directly at the sites of the tombs from the uppermost bed of limestone.

The surfaces of some of the stones from Atteln show evidence of having been in water: they are very smooth, and one stone has a karst groove (Fig. 4.7). This indicates a riverbed provenance. Traces of lime solution, with small drop-like forms, can be seen on some slabs at Etteln (Fig. 4.8). This form of solution indicates a provenance not from a riverbed with flowing water but from an exposed surface that has weathered through the ages.

COMPARISON AND CONCLUSIONS

The geological analyses described above throw light on many details of the character and properties of the quarry sites. Similar circumstances can be assumed for the quarry sites of other gallery graves in Hesse and Westphalia, though these have not yet been studied in detail. A comparison may be drawn between the Züschen tomb, the Altenau valley tombs and the other tombs in respect of the building materials used and the presumed

Fig. 4.8. Traces of lime solution on a slab from Etteln (Photo: Dr M. Hiß)

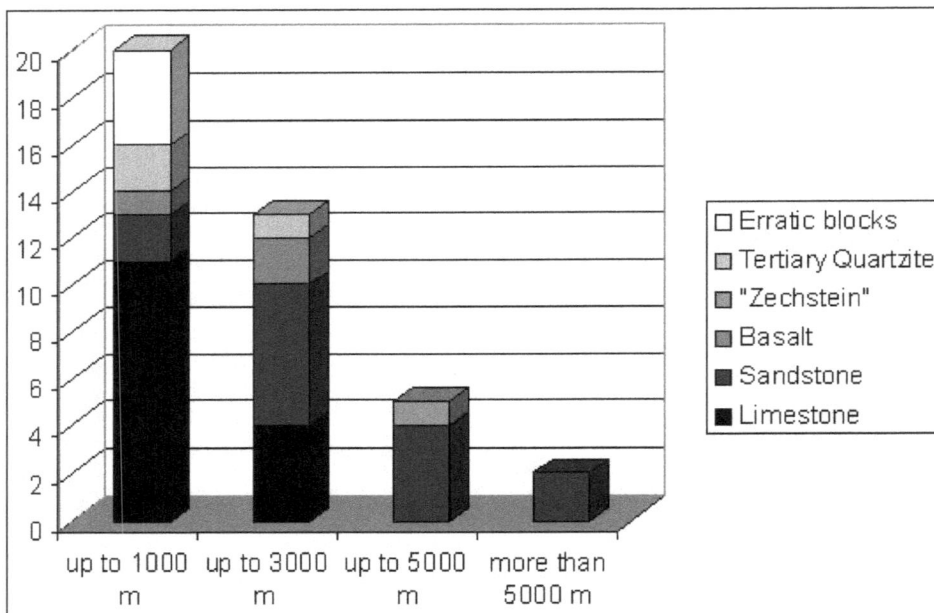

Fig. 4.9. Table of building materials and presumed transport distances for 40 gallery graves
and related tomb types in Hesse, Westphalia and neighbouring areas

distances of transport (Fig. 4.9). It is very likely that the tombs of Kirchborchen and Etteln were built in immediate vicinity to the quarry sites. At Henglarn II a distance of less than 1 km is indicated, though for Atteln I and Züschen I the slabs were transported over greater distances. Limestone and sandstone are the materials employed.

Other gallery graves and related tomb types in Hesse, Westphalia and neighbouring areas are built from erratic blocks, basalt, tertiary quartzite and Permian "Zechstein". Fifty per cent of the tombs are built in close vicinity to the source of the materials, within a distance of up to a kilometre. One third of the tombs incorporated stones from a range up to 3 kilometres. Only seven tombs have a presumed route of transport over longer distances, the two furthest being 11 and 16 kilometres from the sources of their megalithic slabs. It has been demonstrated geologically and must be kept in mind that for all these tombs no other megalithic building materials were available in closer vicinity.

The surfaces of the stones examined in detail in this study show that they come from riverbeds or from close to the surface. These limestone and sandstone beds with their natural propensity to fracture into rectangular and trapezoid slabs are perfect for building, and this must have been evident to the master builders of the Late Neolithic period.

By comparing these results with the provenance of stones from other gallery graves, it is possible to make a distinction in the way the tombs were built. It is evident that the location of some tombs was very probably chosen because of their closeness to the quarry sites. Several other tombs were quite a distance from the sources of their materials, and here it seems that the location had already been chosen for reasons entirely unrelated to the proximity of the stone sources, such as closeness to a settlement, places of special sanctity, or the need to have a megalithic tomb in that specific location.

Acknowledgements

I am grateful to Marion Uckelmann MA for improving my English.

References

BOEHLAU, J.; GILSA, F. VON GILSA ZU (1898) – *Neolithische Denkmäler aus Hessen*. Kassel: Döll.

GÜNTHER, K. (1976) – Zu den neolithischen Stein-kistengräbern von Kirchborchen, Gem. Borchen, Kr. Paderborn. *Germania* 54, 184-191.

GÜNTHER, K. (1978) – Zu den neolithischen Stein-kistengräbern Kirchborchen I und Etteln, Kr. Paderborn. *Germania* 56, 230-233.

GÜNTHER, K. (1979) – Die neolithischen Steinkammer-gräber von Atteln, Kr. Paderborn (Westfalen). *Germania* 57, 153-161.

GÜNTHER, K. (1980) – Die neolithischen Steinkammer-gräber von Henglarn, Kr. Paderborn (Westfalen). *Germania* 58, 147-152.

GÜNTHER, K. (1987) – Ein Großsteingrab in der Warburger Börde bei Hohenwepel, Stadt Warburg, Kreis Höxter. *Ausgrabungen und Funde in Westfalen-Lippe* 4, 65-104.

GÜNTHER, K. (1997a) – *Die Kollektivgräber-Nekropole Warburg I-V*. Mainz: Zabern.

GÜNTHER, K. (1997b) – Das Megalithgrab Atteln I. Ein Grabmal jungsteinzeitlicher Bauern und Viehzüchter im Paderborner Land. *Archäologie in Ostwestfalen* 1, 14-15.

KAPPEL, I. (1989) – Das Grab von Züschen. In *Stein-kammergräber und Menhire in Nordhessen*. (Führer zur nordhessischen Ur- und Frühgeschichte 5) Kassel: Staatliche Kunstsammlungen Kassel, 17-23.

RAETZEL-FABIAN, D. (2000) – *Calden. Erdwerk und Bestattungsplätze des Jungneolithikums. Architektur, Ritual, Chronologie*. Bonn: Habelt.

SCHRICKEL, W. (1966) – *Westeuropäische Elemente im neolithischen Grabbau Mitteldeutschlands und die Galeriegräber Westdeutschlands und ihre Inventare*. Bonn: Habelt.

SCHWELLNUS, W. (1979) – *Wartberggruppe und hessische Megalithik. Ein Beitrag zum späten Neolithikum des Hessischen Berglandes*. Wiesbaden: Landesamt für Denkmalpflege Hessen.

BEYOND STONEHENGE:
SEEKING THE START OF THE BLUESTONE TRAIL

Timothy DARVILL

Centre for Archaeology, Anthropology and Heritage, School of Conservation Sciences,
Bournemouth University, Fern Barrow, Poole, Dorset, BH12 5BB

Abstract: This paper explores the arrangement of stones at Stonehenge in Wiltshire. It focuses on the disposition of rhyolites, volcanic tuffs, and dolerites, collectively known as the 'bluestones', that were transported to the site from the Preseli Hills of southwest Wales some 250km away to the west. It is shown that the way in which the stones were erected at Stonehenge corresponds closely with their original natural arrangement in Wales. It is proposed that the blocks of spotted dolerite used in building the central structure at Stonehenge had some special meaning to those who used the site; folklore evidence suggests this may be connected with perceived healing properties.
Key words: Stonehenge; Neolithic Britain; stone circle; henge; Bluestone; Preseli Hills

Résumé: Dans cet article nous examinons la disposition des pierres de Stonehenge dans le Wiltshire. L'étude se focalise sur la distribution spatiale des rhyolites, des tufs volcaniques et des dolérites communément appelés les 'bluestones' et qui ont été transportés depuis le site des collines de Preseli situé à quelques 250 km à l'ouest. On remarque que la façon avec laquelle les pierres ont été levées à Stonehenge est étroitement liée à leur répartition d'origine naturelle au Pays de Galles. On propose que les blocs en dolérite tachetée employés dans la construction de la structure centrale de Stonehenge possédaient une signification particulière pour les gens qui fréquentaient le site. Les témoins folkloriques nous amènent à suggérer que cette signification peut avoir été mise en relation avec des propriétés curatives.
Mots-clés: Stonehenge, Néolithique britannique, cercles de pierres dressées, Bluestones, collines de Preseli

Stonehenge on the southern edge of Salisbury Plain in central southern England is probably the best-known and most-visited megalithic monument in Europe (Fig. 5.1). Iconic in its form, it was clearly an elaborate structure that Richard Atkinson argued involved the application of technologies from carpentry to the formation of an enduring stone edifice (1979, 177). As long ago as 1740, William Stukeley recognized that the builders of Stonehenge used several different kinds of stone in its construction (Stukeley 1740, 5). It was a matter that became a topic for comment over the following centuries, but it was not until the late nineteenth century that the relatively new science of petrology began to add precision to the sourcing of the stones, and it was not until the early twentieth century, and the work of H.H. Thomas, that the detail become widely known (Thomas 1923).

Leaving aside the small portable objects found at and around the monument, two major and two minor sources are represented (Darvill 2006, table B for summary). The minor ones comprise pieces of Old Red Sandstone, the largest of which was used for the so-called Altar Stone in the centre of the monument, and oolitic limestone, whose structural use is uncertain. First among the major components are the 80 or so blocks of hard siliceous sandstone, known locally as 'sarsen'. These derive from a variety of sources on the chalk downs of central southern England within about 50km of Stonehenge (Bowen & Smith 1977; Howard in Pitts 1982, 119-23). They are the most visually dominant elements of the monument, the largest weighing about 40 tons. The second major source is represented by a group of at least 82 blocks know as the 'bluestones' which represent a heterogeneous collection of dolerites, rhyolites, and volcanic tuffs whose minera-

logy and chemistry can be matched with surface outcrops along the Preseli Hills of southwest Wales some 250km west of Stonehenge as the crow flies (Thomas 1923; Thorpe *et al.* 1991; Green 1995). They vary considerably in size and shape, the largest weighing about 4 tons, one tenth of the size of the largest sarsen.

Archaeological investigations at Stonehenge during the twentieth century showed that what is visible today at the site is the decayed remains of several phases of construction and use spanning a period of more than 1500 years from around 3000 BC through to 1500 BC. The earliest elements comprised a circular earthwork (bank and external ditch) within which were timber structures of various kinds, and a large number of cremation burials (Cleal *et al.* 1995, 63-165). Around 2300 BC the first stone structures were built: a double circle of bluestones from southwest Wales closely associated with Beaker style pottery (Cleal *et al.* 1995, 169-88; and see Case 1997). One of the most significant things about Stonehenge is the fact that once these bluestones arrived at the site they were incorporated into all the subsequent recognizable changes and re-modellings. This is what makes Stonehenge special, and it is these components that are the focus of this paper. Let us first look in a little more detail at the bluestones at Stonehenge, then turn to the bluestone outcrops in the Preseli Hills, and finally bring some of the findings together within a more broadly-based social interpretation.

BLUESTONES AT STONEHENGE

Little is known about the first bluestone structure at Stonehenge, as it was subsequently demolished. Its

Fig. 5.1. Stonehenge from the northeast looking southwestwards. (Photo: Timothy Darvill)

footprint can be seen in the dumb-bell shaped pairs of stone sockets (the Q and R holes) revealed by excavations, but not all of its circuit has ever been explored and for this reason it is sometimes depicted as a rather odd structure on reconstruction drawings. Assuming the available excavated sections are representative, and there is no reason to think otherwise, then the first bluestone monument must be seen as a double circle of fairly close-set bluestone pillars, about 26 metres across, with an entrance aligned on the midsummer sunrise to the northeast (Fig. 5.2, Phase 3i). Some of the pillars, perhaps those flanking the entrance, might have supported lintels to form small trilithons, since two such components have been found re-used in later contexts. At least 82 stones would have been required for this structure. There may also have been internal features, most probably the Altar Stone and a grave near the entrance.

This double bluestone circle was dismantled sometime around 2200 BC during a major re-modelling which resulted in the development of the structure still visible today (Cleal *et al.* 1995, 188-264). In this, the bluestones seem to have been re-used to form a pair of concentric rings (Fig. 5.2, Phase 3iv), widely spaced with the great sarsen trilithons set in the gap between. Whether the original order and arrangement of bluestones was retained during this re-modelling we do not know, neither it is known whether any additional bluestones were brought

from southwest Wales at this time. As set out in the re-modelled monument, however, there are some interesting and potentially significant patterns.

The outer bluestone ring was roughly circular and originally comprised about 40 stones. Just over half of them remain standing, and all of them are seemingly natural unshaped blocks in a wide variety of lithologies including dolerites, rhyolites, and tuffs from the Sealyham and Fishguard volcanic groups (Fig. 5.3A). To what extent they were selected for their shape and form is hard to say, but variations in colour and texture are visible to the naked eye notwithstanding the surface patina that has built up during 4000 years of exposure to the elements. It may also be noted that the stones flanking the northeastern entrance each have a distinctive profile: a pointed one on the left and a flat-topped one on the right. This follows a widespread and long-lived pattern seen in British megaliths, and can be linked to a binary sexual symbolism, male to the left and female to the right when viewed from inside the circle looking outwards (Darvill 2004a, 51-2).

The inner bluestone ring is oval in plan and originally comprised about 25 stones. Later in its life five or six pillars were removed from the northeastern sector to create a bluestone horseshoe defining the central area of the monument (Fig. 5.2, Phase 3v). Thirteen stones remain, all of them a distinctive kind of spotted dolerite.

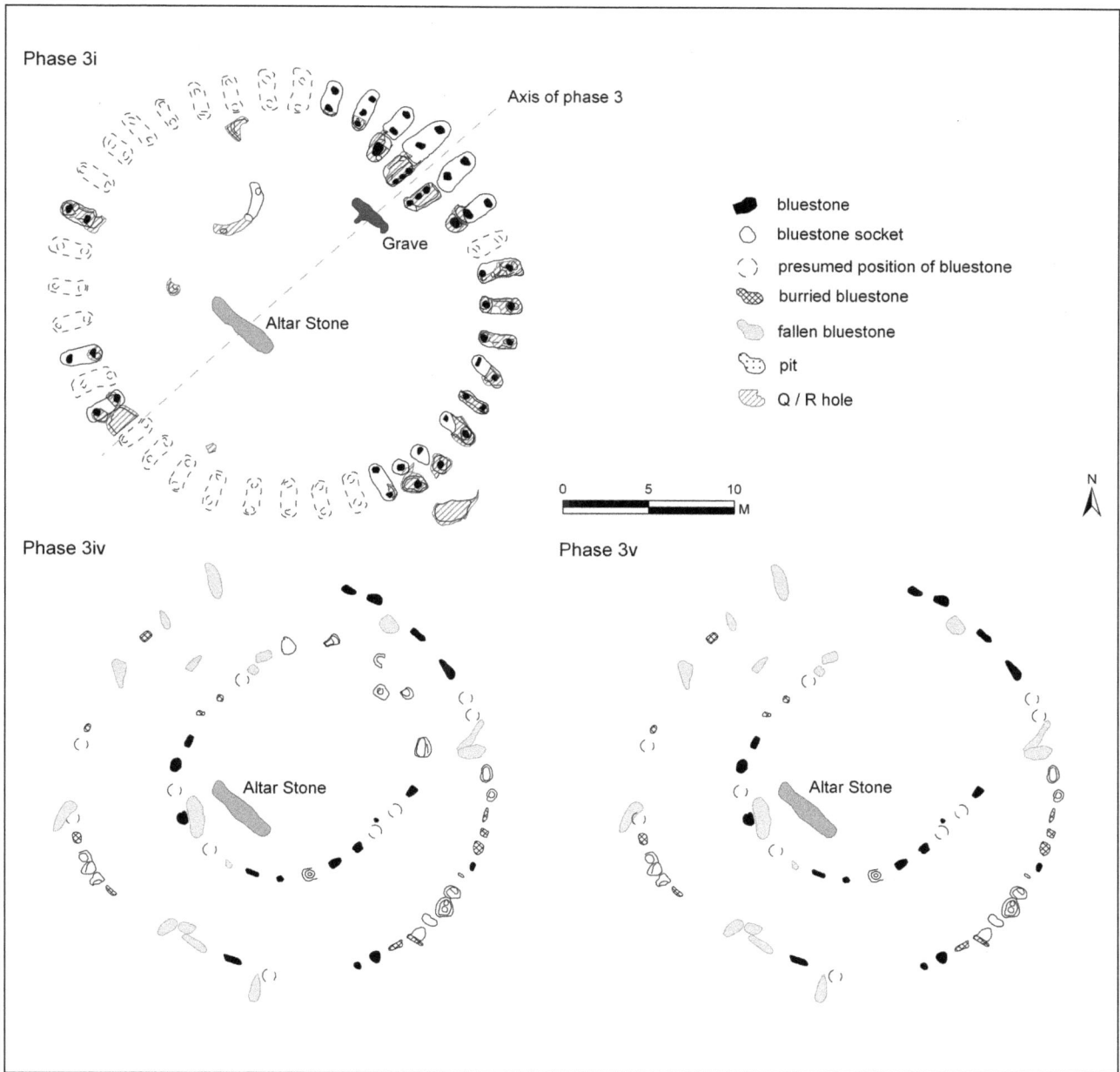

Fig. 5.2. Phase plans showing the arrangement of bluestones at Stonehenge during successive re-modellings between c.2600 and 2200 BC. (Drawing created by Vanessa Constant after Cleal *et al.* 1995, figs 80, 81,116, and 117)

To the naked eye they all look similar and have the same surface characteristics and colour. Four have been sampled for chemical analysis by geologists from the Open University and all of these come from a restricted set of outcrops at or near Carn Menyn in the very heart of the Preseli Hills (Thorpe *et al.* 1991, 126; Green 1995). Visual inspection confirms that the unsampled pillars are of the same spotted dolerite. In contrast to the pillars of the outer bluestone circle, this inner structure comprises artificially shaped blocks with fairly regular parallel-sided forms that reflect the highly columnar nature of the stone at Carn Menyn (Fig. 5.3B). There is no known rock art on them, but the stones are graded in height with the tallest to the southwest.

BLUESTONES IN PRESELI

Despite the fact that the source of the Stonehenge bluestones has been known for almost a century, very little co-ordinated research has been done in and around the Preseli Hills to investigate the start of the bluestone trail that eventually leads to Stonehenge. Geological investigations have usefully emphasised the wide range of rock types present within a geologically complicated landscape (Evans 1945; Bevins *et al.* 1989), but much of the work linking the rock outcrops with archaeological structures has been directed towards 'proving' that glacial action was responsible for moving these rocks from southwest Wales to Salisbury Plain (Thorpe *et al.* 1991;

A   B

Fig. 5.3. Bluestones at Stonehenge (Phase 3v). A: Stone 46 in the outer Bluestone Circle.
B: Stone 61 in the inner Bluestone Horseshoe. (Photos: Timothy Darvill)

Williams-Thorpe *et al.* 1997; 2004; 2006), an impossibi-lity given the complete absence of any suitable eastward flowing glaciers and a dead-end research path (Scourse 1997; Castleden 2001). As H.H. Thomas long ago concluded, quoting the well-known Salisbury antiquarian Frank Stevens, 'alluring as the Glacial Drift Theory may appear, it must reluctantly be set aside for want of convincing evidence' (Thomas 1923, 252).

Equally, however, archaeologists can be criticised for failing to contextualize the geological work in relation to the treatment and use of particular kinds of stone, and for not exploring the archaeology of the Preseli landscape in order to provide a broader cultural background. Exceptions include the investigations carried out by W.F. Grimes in the late 1920s and 30s (Grimes 1929; 1938; 1939a; 1939b), and the walk-over surveys and plotting directed by Peter Drewett in the early 1980s (Drewett 1983; 1984; 1985), but in general these were small-scale operations and have never been fully published. The need to move beyond Stonehenge and examine the archaeology of the Preseli Hills in detail lies at the heart of a project initiated in 2001 by the present author and Geoffrey Wainwright, in association with the Royal Commission on the Ancient and Historical Monuments of Wales. Called SPACES – the Strumble-Preseli Ancient Communities and Environment Study – this project is using a combination of remote sensing, topographic survey, and selected excavation to establish the archaeology of a broad swathe of land from the coastal fringe right through the Preseli uplands to their eastern extremity near Crymmch. Already the results have been impressive (Darvill & Wainwright 2002a; 2002b; Darvill *et al.* 2003; 2004; 2005).

Comparing the geological attribution of the bluestones as they are set up at Stonehenge with the actual arrangement of outcrops on the ground in west Wales there is a close spatial correspondence (Fig. 5.4; and see Darvill 2006, 136-41). The range of dolerites, rhyolites and tuffs in the outer circle derive from a scatter of rock sources which surround the outcrops of spotted dolerite at Carn Menyn that were used for the inner oval/horseshoe at Stonehenge. In this respect Stonehenge seems to have been built as a representation of a reality that exists in the structure and arrangement of a particular and very real place; the translation across space of some fragment of a world that must have had special meaning for those who conceived and undertook the task of physically moving the earth.

Some of the rhyolites and tuffs may have been quarried from the ground in a fairly conventional way, and finding those quarries is an objective for future work. The spotted dolerite is rather different. Blocks of this material naturally fracture from the sides of exposed rock faces and litter the landscape (Fig. 5.5). Quarrying here simply

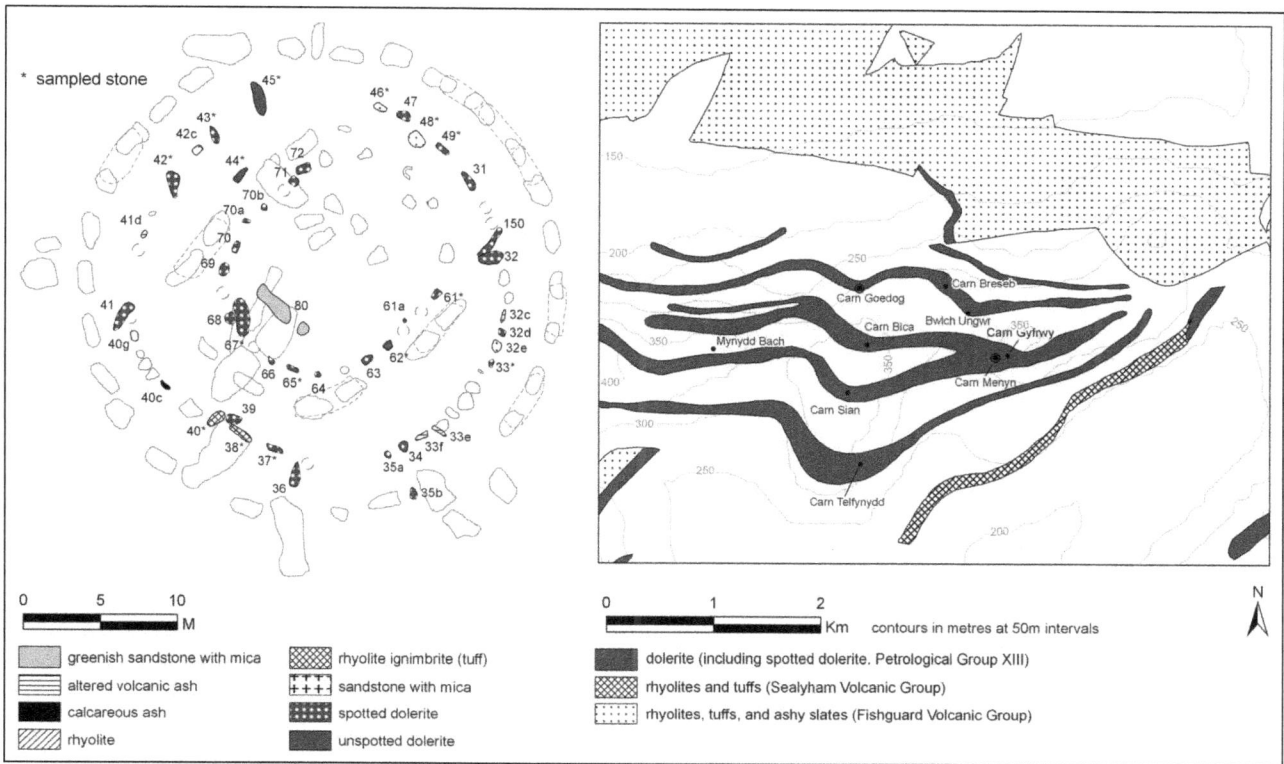

Fig. 5.4. Arrangement of the bluestones in Phase 3v at Stonehenge (left) and principal rock outcrops in the eastern Preselis (right). (Drawing created by Vanessa Constant after Darvill 2006, fig. 50)

Fig. 5.5. Outcrops of spotted dolerite on Carn Menyn, Pembrokeshire. (Photo: Timothy Darvill)

means prising a suitable desired block from the ground or lifting it from a spread of clitter. The best blocks are those on Carn Menyn itself, the highest point in this part of the Preselis. Detailed surveys here have revealed the presence of a stone wall built in a style typical of fourth and third millennia BC enclosures in the west of Britain whereby natural outcrops were linked together by walling. Small-scale excavations in 2005 showed the neat construction of the wall and the possible presence of an outer ditch. The lack of datable material is perhaps to be expected at a structure whose purpose is to define and bound a series of special rock outcrops, but environmental information from the old ground surface below the bank ties it neatly with the cleared upland landscapes of the third millennium BC known from dated sequences elsewhere in the area (Darvill *et al.* 2005, 17-22). Within and around the Carn Menyn enclosure there are numerous pillar stones in various states of preparation, some broken. There are also scatters of working debris indicative of primary shaping. Independent dating is impossible, but there is a close correspondence between the shape, size, and working of the pillars remaining around Carn Menyn and those that were set up at Stonehenge.

Space does not permit a full account of the wealth of archaeology known along the Preseli Hills as a result of SPACES, but especially noteworthy is the stone oval at Bedd Arthur situated just 1500 metres west of Carn Menyn, overlooking the outcrops themselves, and in plan and orientation a very close match indeed for the central oval at Stonehenge (Darvill & Wainwright 2003, fig. 15). Indeed, SPACES has shown very clearly that many of the monument classes of fourth and third millennium BC date familiar in Wessex are also represented in southwest Wales.

STONEHENGE AND THE BLUESTONE SOURCES

Stone was clearly a highly significant material to people of the fifth through to the third millennium BC in northwest Europe. Small pieces in the form of axes and ornaments travelled very considerable distances indeed (Clough & Cummins 1988), and larger blocks were seemingly moved in the course of monument building on a scale that is slowly being appreciated (cf. Thorpe & Williams-Thorpe 1991; Burl 1991a). Stonehenge is by no means the only megalithic monument in Britain to incorporate a range of materials from both local and distant sources. Stones for the great passage grave of Newgrange in the Boyne Valley of Ireland came from Slieve Croob more than 50km to the north and the Wicklow Mountains nearly double that distance to the south (Mitchell 1992). The West Kennet long barrow near Avebury in Wiltshire includes Corallian limestone and Forest Marble limestone from outcrops between 10km and 30km away to the west (Piggott 1962, 14), and similar traditions are represented at other Cotswold-Severn long barrows (Darvill 2004b, 114). The passage grave of La Hougue Bie on Jersey includes orthostats from at least seven sources in the eastern part of the Island, some more than 7km away (Patton 1992, 393), while the four or five great blocks of millstone grit each weighing about 30 tonnes that were used to make the Devil's Arrows beside the River Ure in North Yorkshire came from Plumpton Rocks near Knaresborough some 14km to the south (Burl 1991b). It is fair to say that the Stonehenge bluestones are rather exceptional in having been moved a distance of more than 250km, but this is extreme rather than anomalous. As Richard Bradley and others have noted, the material that was moved the furthest may be regarded as the most 'special' and seems to derive from remote sources near the very tops of mountains that today we might consider difficult of access and uneconomic to exploit (Bradley 2000, 81-90). Of course, in a prehistoric context such considerations may well have been quite alien, and even basic categories that we use today in order to describe seemingly simple materials – wood and stone for example – may be far more problematic than we imagine. Can we even assume that such distinctions applied?

It might be argued that a sacred geography in which places and areas were imbued with a significance related to wider cosmological schemes can also include a dynamic or living element whereby pieces of a place can be removed to another context in order transmit those same meanings or stimulate memories of them. In this sense the representational dimensions of the material itself link the place of origin with the place of use, but they leave the problem of defining what the significance, symbolic or actual, of these translated blocks of stone might be (see Insoll 2006 for the use of powerful stones in 'franchising' shrines). It is hard to imagine that it was simply the intrinsic nature of the stone itself that was of interest, however beautiful such stone might look to us now. Colour may have been important, especially if the colour of stone in the ground was somehow connected to ideas about creation and linked particular myths and legends. The presence of white inclusions in the spotted dolerite from Preseli may be somehow relevant; we might speculate that these were tears wept by the ancients, raindrops from the storm that brought the world into being, or the stars of heaven mirrored in the earth beneath our feet: but now we are guessing in the dark. Acoustic properties; links with special events; or simply a piece of the mountain that was home to the gods may be equally plausible – but are also equally speculative.

Could folklore and tradition help? Local place-names provide no real insights. 'Carn Menyn', first recorded in the sixteenth century AD, simply means the 'butter hill' and probably refers to butter-making in the area at that time (Charles 1992, 125). The name 'Preseli' can be traced back to the thirteenth century AD and seemingly derives from a personal name meaning 'Selyf's thicket' (Charles 1992, 26). More rewarding, however, is the work of Geoffrey of Monmouth, a twelfth century cleric and author of an influential work entitled the *History of the Kings of Britain* (see Grinsell 1975). Woven through the text is a folk-tale, perhaps of Welsh or Breton origin, which records that the stones for building Stonehenge were brought to Salisbury Plain by Merlin the wizard from a mountain named as 'Killaraus' situated in western Britain (Hibernia/Ireland is specifically mentioned) and were selected for the construction of a memorial to those lost in a great battle because they had curative powers (Thorpe (trans.) 1966, 196). Stuart Piggott and others have accepted that Geoffrey's folk-tale drew on oral traditions whose roots lay deep in prehistory, and that we should pay some heed to what he says (Piggott 1941). But not everyone agrees. Aubrey Burl, for example, declared that Geoffrey's accounts were 'no more than a monkish mixture of Merlin, magic and imagination' and places as much faith in his 'history' as we might in our own unsupported speculations (Burl 1985, 182).

It is a fact, however, that a number of so-called holy wells or healing-wells exist in the form of springs issuing from the sides of the Preselis (Jones 1992; Darvill *et al.* 2004, 106-8). A belief that curative powers resided in the stone from which such waters came would not only provide a very powerful reason for transporting pieces of the mountain to a distant land, but also perhaps gave purpose to the structure that became Stonehenge, and a reason for people to go there for celebrations and festivals.

Acknowledgements

Thanks to Geoff Wainwright for many fruitful discussions about Stonehenge and its purpose, to Yvette Staelens for help preparing this paper, and to Vanessa Constant for preparing the illustrations.

References

ATKINSON, R.J.C. (1979) *Stonehenge* (Revised edition). Harmondsworth: Penguin.

BEVINS, R.E.; LEES, G.J.; ROACH, R.A. (1989) Ordovician intrusions of the Strumble Head – Mynydd Preseli region, Wales: lateral extensions of the Fishguard volcanic complex. *Journal of the Geological Society of London* 146, 113-23.

BOWEN, C., SMITH, I.F. (1977) Sarsen stones in Wessex: the Society's first investigations in the evolution of the landscape project. *Antiquaries Journal* 57, 186-96.

BRADLEY, R. (2000) *An archaeology of natural places.* London: Routledge.

BURL, A. (1985) Geoffrey of Monmouth and the Stonehenge bluestones. *Wiltshire Archaeological and Natural History Magazine* 79, 178-83.

BURL, A. (1991a) Megalithic myth or man the mover? *Antiquity* 65, 297-8

BURL, A. (1991b). The Devil's Arrows: the archaeology of a stone row. *Yorkshire Archaeological Journal* 63, 1-24.

CASE, H.J. (1997) Stonehenge revisited. A review article. *Wiltshire Archaeological and Natural History Magazine* 90 161-8.

CASTLEDEN, R. (2001) The epic of the Stonehenge bluestones: were they moved by ice, or by people? *3rd Stone* 39, 12-25.

CHARLES, B.G. (1992) *The place-names of Pembrokeshire.* Aberystwyth: The National Library of Wales (2 volumes).

CLEAL, R.; WALKER, K.E.; MONTAGUE, R. (1995) *Stonehenge in its landscape. Twentieth-century excavations.* London: English Heritage (English Heritage Archaeological Report 10).

CLOUGH, T.H.McK; CUMMINS, W.A. (eds.) (1988) *Stone Axe Studies. Volume 2. The petrology of prehistoric stone implements from the British Isles.* London: Council for British Archaeology (Council for British Archaeology Research Report 67).

DARVILL, T. (2004a) Tales of the land, tales of the sea: people and presence in the Neolithic of Man and beyond. In Cummings, V.; Fowler, C. eds., *The Neolithic of the Irish Sea: materiality and traditions of practice.* Oxford: Oxbow Books, 46-54.

DARVILL, T. (2004b) *Long barrows of the Cotswolds and surrounding areas.* Stroud: Tempus.

DARVILL, T. (2006) *Stonehenge: the biography of a landscape.* Stroud: Tempus.

DARVILL, T.; MORGAN EVANS, D.; WAINWRIGHT, G. (2003) Strumble-Preseli Ancient Communities and Environment Study (SPACES): Second report 2003. *Archaeology in Wales* 43, 3-12.

DARVILL, T.; MORGAN EVANS, D.; WAINWRIGHT, G. (2004) Strumble-Preseli Ancient Communities and Environment Study (SPACES): Third report 2004. *Archaeology in Wales* 44, 104-8.

DARVILL, T.; MORGAN EVANS, D.; FYFE, R.; WAINWRIGHT, G. (2005) Strumble-Preseli Ancient Communities and Environment Study (SPACES): Fourth report 2005. *Archaeology in Wales* 45, 17-24.

DARVILL, T.; WAINWRIGHT, G. (2002a) SPACES – exploring Neolithic landscapes in the Strumble-Preseli area of southwest Wales. *Antiquity* 76, 623-4.

DARVILL, T.; WAINWRIGHT, G. (2002b) Strumble-Preseli Ancient Communities and Environment Study (SPACES): First Report 2002. *Archaeology in Wales* 42, 17-28.

DARVILL, T.; WAINWRIGHT, G. (2003). Stone circles, oval settings and henges in southwest Wales and beyond. *Antiquaries Journal* 83, 9-45.

DREWETT, P. (1983) *Mynydd Preseli 1983. First interim report.* London: Institute of Archaeology, University of London (limited circulation typescript report).

DREWETT, P. (1984) *Mynydd Preseli 1984. Second interim report.* London: Institute of Archaeology, University of London (limited circulation typescript report).

DREWETT, P. (1985) *Mynydd Preseli 1985. Third interim report.* London: Institute of Archaeology, University of London (limited circulation typescript report).

EVANS, W.D. (1945) The geology of the Prescelly Hills, North Pembrokeshire. *Quarterly Journal of the Geological Society of London* 64, 273-96.

GREEN, C.P. (1997). The provenance of rocks used in the construction of Stonehenge. In Cunliffe, B; Renfrew, C. eds., *Science and Stonehenge*. Oxford: The British Academy (Proceedings of the British Academy; 92), 257-70.

GRIMES, W.F. (1929) Pembrokeshire survey. *Bulletin of the Board of Celtic Studies* 5, 277.

GRIMES, W.F. (1938) Excavations at Meini Gwyr, Carmarthen. *Proceedings of the Prehistoric Society* 4, 324-25.

GRIMES, W.F. (1939a) Meini Gwyr, Carmarthenshire. *Bulletin of the Board of Celtic Studies* 9, 373-74.

GRIMES, W.F. (1939b) Bedd y Afanc. *Proceedings of the Prehistoric Society* 5, 258.

GRINSELL, L.V. (1975) *Legendary history and folklore of Stonehenge*. St Peter Port: Toucan Press.

INSOLL, T. (2006) Shrine franchising and the Neolithic in the British Isles: some observations based upon the Tallensi, Northern Ghana. *Cambridge Archaeological Journal* 16, 223-238.

JONES, F. (1992) *The Holy Wells of Wales*. Cardiff: University of Wales Press.

MITCHELL, G.F. (1992) Notes on some non-local cobbles at the entrance to the passage-graves at Newgrange and Knowth, county Meath. *Journal of the Royal Society of Antiquaries of Ireland* 122, 128-45.

PATTON, M. (1992) Megalithic transport and territorial markers: evidence from the Channel Islands. *Antiquity* 66, 392-5.

PIGGOTT, S. (1941) The sources of Geoffrey of Monmouth. II. The Stonehenge story. *Antiquity* 15, 305-19.

PIGGOTT, S. (1962) *The West Kennet Long Barrow excavations 1955-56*. London: HMSO (Ministry of Works Archaeological Reports 4).

PITTS, M. (1981) On the road to Stonehenge: Report on investigations beside the A344 in 1968, 1979 and 1980. *Proceedings of the Prehistoric Society* 48, 75-132.

SCOURSE, J.D. (1997) Transport of the Stonehenge bluestones: testing the glacial hypothesis. In Cunliffe, B; Renfrew, C. eds., *Science and Stonehenge*. Oxford: The British Academy (Proceedings of the British Academy 92), 271-314.

STUKELEY, W. (1740) *Stonehenge, a temple restor'd to the British druids*. London: W Innys & R Manby.

THOMAS, H.H. (1923) The source of the stones of Stonehenge. *Antiquaries Journal* 3, 239-60.

THORPE, L. (trans.) (1966) *Geoffrey of Monmouth. The history of the Kings of Britain*. Harmondsworth: Penguin.

THORPE, R.S.; WILLIAMS-THORPE, O. (1991) The myth of long-distance megalithic transport. *Antiquity* 65, 64-73.

THORPE, R.S.; WILLIAMS-THORPE, O.; JENKINS, D.G.; WATSON, J. (1991) The geological sources and transport of the bluestones of Stonehenge, Wiltshire, UK. *Proceedings of the Prehistoric Society* 57, 103-57.

WILLIAMS-THORPE, O.; GREEN, C.P.; SCOURSE, J.D. (1997) The Stonehenge bluestones: a discussion. In Cunliffe, B; Renfrew, C. eds., *Science and Stonehenge*. Oxford: The British Academy (Proceedings of the British Academy 92), 315-34.

WILLIAMS-THORPE, O.; POTTS, P.J.; JONES, M.C. (2004) Non-destructive provenancing of bluestone axe-heads in Britain. *Antiquity* 78, 359-79.

WILLIAMS-THORPE, O.; JONES, M.C.; POTTS, P.J.; WEBB, P. C. (2006) Preseli dolerite bluestones: axe-heads, Stonehenge monoliths, and outcrop sources. *Oxford Journal of Archaeology* 25, 29-46.

ARCHITECTONIQUE ET ESTHETIQUE DES ALIGNEMENTS DE MENHIRS DU SUD DE LA VENDEE (FRANCE)

Gérard BENETEAU-DOUILLARD

16 rue du Logis, 85320 La Bretonnière, France

Abstract: The stone rows of the southern Vendée stand out among the megalithic monuments of the Atlantic façade of western France in possessing a gendered anthropomorphism. This is defined by the intentional selection of the blocks that were to be extracted and by the way those blocks were then roughly shaped through pounding in order to accentuate their distinctively anthropomorphic features. The method of erection, traces of which have been discovered during excavation, demonstrates that the cradle used for the transport of the blocks also played a part in the process of their erection. It defined the orientation of the faces of the menhirs, and hence produced a repeated regularity in the way the stones are positioned within their sockets.
Key words: menhirs, architectonic, aesthetic, extraction, erection

Résumé: Unité architecturale du mégalithisme de la façade atlantique de l'ouest de la France, les alignements de menhirs du sud de la Vendée ont l'originalité de posséder un anthropomorphisme sexualisé. Celui-ci est défini par un choix prédéterminé des rochers à extraire et par une régularisation sommaire de la roche par martelage, qui accentuent les parties anthropomorphes évocatrices. Les modalités de levage, dont les traces ont été retrouvées à la fouille, montrent que l'armature de transport des blocs participait au mécanisme d'érection, définissait l'orientation des faces des menhirs et entraînait ainsi la reproduction similaire de leur ancrage dans les fosses.
Mots clé: menhirs, architectonique, esthétique, extraction, levage

Bien que très présents dans tout l'ouest de la France (et au-delà!), les pierres debout et autres menhirs, isolés, dressés par paires ou en petits groupes, n'ont jamais fait l'objet de travaux spécifiques, en dehors des grands alignements carnacéens comptant généralement plusieurs centaines de pierres (Lecerf 1999).

TYPOLOGIE ET ORIENTATION

Les alignements courts du sud de la Vendée (communes de St Hilaire la Forêt, Avrillé, Le Bernard et Le Givre, pour l'essentiel), étudiés de 1984 à 2003, ont permis la détermination d'une typologie architecturale, associant un nombre récurrent et précis de pierres, avec la présence d'un ou plusieurs menhirs anthropomorphes (Benéteau 1999, 170-3). Les hauteurs de ces grandes pierres atteignent 3 à 7 mètres.

1) Les alignements en "Frontispice" ou en "Façade", comprenant toujours un nombre impair de pierres (de 3 à 7 connues), accompagnant un menhir "géant" placé au centre de la ligne. Celui-ci détermine l'axe d'une esthétique monumentale dont la symétrie, manifestement préconçue et recherchée, est engendrée par un effet d'élévation graduelle des menhirs latéraux (Fig. 6.2).

2) Les alignements en "Cortège", qui possèdent un nombre pair de monolithes (toujours quatre, d'ailleurs), associant, là aussi, une pierre "géante", invariablement élevée en extrémité de ligne, et une file de pierres (trois), beaucoup plus petites. La première de ces pierres est souvent accolée au menhir "géant".

3) Les alignements à "Satellite", dont l'unique spécimen jusqu'à présent déterminé avec certitude semble com-

biner les deux précédentes architectures. Il se peut, pour ce cas, que l'on soit en présence d'un monument "spécifique".

Ces files de menhirs sont toutes élevées selon un axe nord-sud (en fait, selon un cadran qui va du nord-nord-ouest, sud-sud-est, au nord-nord-est, sud-sud-ouest). Toutes les faces plates (ou reconnues comme telles), sont orientées à l'est, et les faces bombées (ou irrégulières) à l'ouest. Cette permanence architecturale n'est pas fortuite, et la similarité du blocage des monolithes dans les fosses conforte l'existence plausible de règles de construction.

ORIGINE ET CHOIX DES BLOCS

La presque totalité des blocs composant ces alignements sont issus des affleurements granitiques locaux, dont les bancs d'extraction peuvent se situer de quelques dizaines de mètres à près d'un kilomètre de distance (et même plusieurs kilomètres pour deux cas repérés). La position de certaines de ces carrières, ouvertes sur des pentes, laisse imaginer l'importance et l'audace des efforts déployés pour déplacer les plus gros monolithes jusqu'aux sommets des élévations de terrain, sur lesquels ils sont toujours érigés. Il existe aussi quelques rares blocs de provenance exceptionnelle, en grès, qui ont fait l'objet d'un transport pouvant atteindre trois kilomètres, et dont la présence au sein de quelques alignements, revêt un caractère symbolique territorial probable (Benéteau 1999, 191). L'étude de la morphologie des blocs composants ces alignements, le repérage des carrières et affleurements possédant encore des mégalithes en cours d'extraction, ont permis de déterminer les choix qui ont guidé les mégalitheurs.

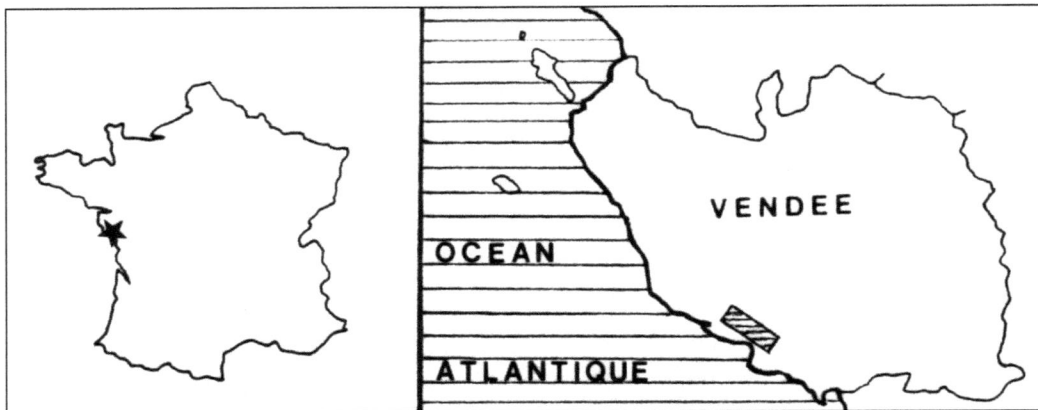

Fig. 6.1. Situation géographique de la zone étudiée

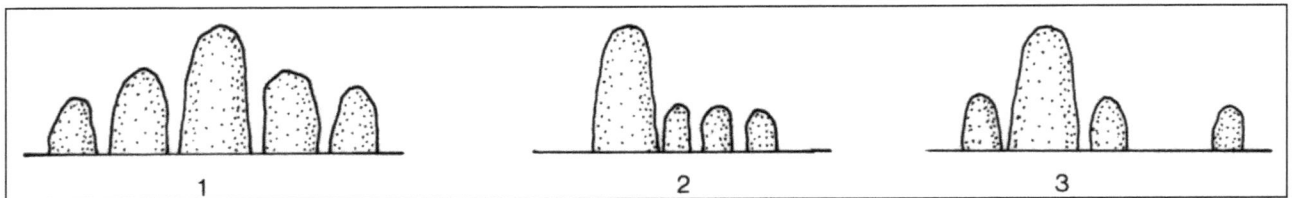

Fig. 6.2. Corpus des différents types d'alignements courts (d'après les fouilles)

En effet, l'observation des blocs en cours d'arrachement sur la matrice rocheuse fait pressentir la volonté d'une détermination des formes, évoquant avec constance une silhouette anthropomorphe "naturelle". Certains affleurements semblent même avoir bénéficiés de travaux de régularisation de leurs reliefs originels, sur place, avant arrachement (Fig. 6.3b).Ces régularisations paraissent également avoir été précédées d'un traçage des formes voulues, sur la roche, elles-mêmes "orientées" esthétiquement par la morphologie des blocs sélectionnés. Certains sillons, plus fins, jouxtant les rainures de débitage, pourraient être interprétées en ce sens. La recherche presque "obsédante", de la silhouette humaine sur les émergences rocheuses, pourrait faire qualifier "l'ambiance" sociale de ces populations, d'animiste.

De plus, l'existence de cupules (parfois en unique spécimen, mais aussi en compositions de deux à cinq cupules), à la surface de ces formes choisies, se retrouve également sur les menhirs érigés (parfois elles sont presque effacées par l'érosion post-mégalithique); indifféremment sur les faces d'affleurements comme sur les faces d'arrachement. La problématique générale des cupules ne sera pas traitée ici, mais force est de constater (et le cas de cette région n'est pas isolé), qu'il semble bien avoir une relation étroite entre le choix des blocs et la présence de ce genre de gravure.

S'agit-il d'un "marquage" de la roche; opération précédant l'extraction?

Bien évidemment, ces cupules en petit nombre peuvent être difficilement comparées avec les grandes com-binaisons iconographiques cupulaires, pouvant être associées à d'autres types de gravures (Benéteau à paraître). Mais si les mégalitheurs semblent bel et bien avoir créés et développés de véritables carrières, au sein desquelles parfois, de nombreux volumes importants furent extraits, ils jouèrent également la carte de l'opportunisme en utilisant des masses dont les formes et les prédispositions naturelles de débitage favorisaient leurs critères de références. C'est le cas de "l'ovoïde de La Pierre" (Fig. 6.3a), qui se présentait dans son état géologique originel comme un gros œuf légèrement aplati, dont le volume était naturellement partagé en son milieu, par un filon d'aplite. Creusant la diaclase bordant le filon, les préhistoriques "décollèrent" les deux parties et les érigèrent en menhirs, de part et d'autre d'un grand monolithe, haut de cinq mètres, qu'ils dressaient au centre de l'alignement.

Mais là ne s'arrêta pas ce pragmatisme de circonstance. S'apercevant que les deux "moitiés" de l'ovoïde étaient opposables, les constructeurs les plantèrent:

1) les faces plates côté est,

2) de façon à ce que la plus grande arête de chaque bloc soit placée du côté du grand menhir central, accentuant encore l'effet d'anthropomorphisme à l'ensemble de l'alignement (Benéteau 1999, 174).

Il s'agit là, d'un véritable "cas d'école" dans l'architectonique employée pour la construction de ces alignements de menhirs.

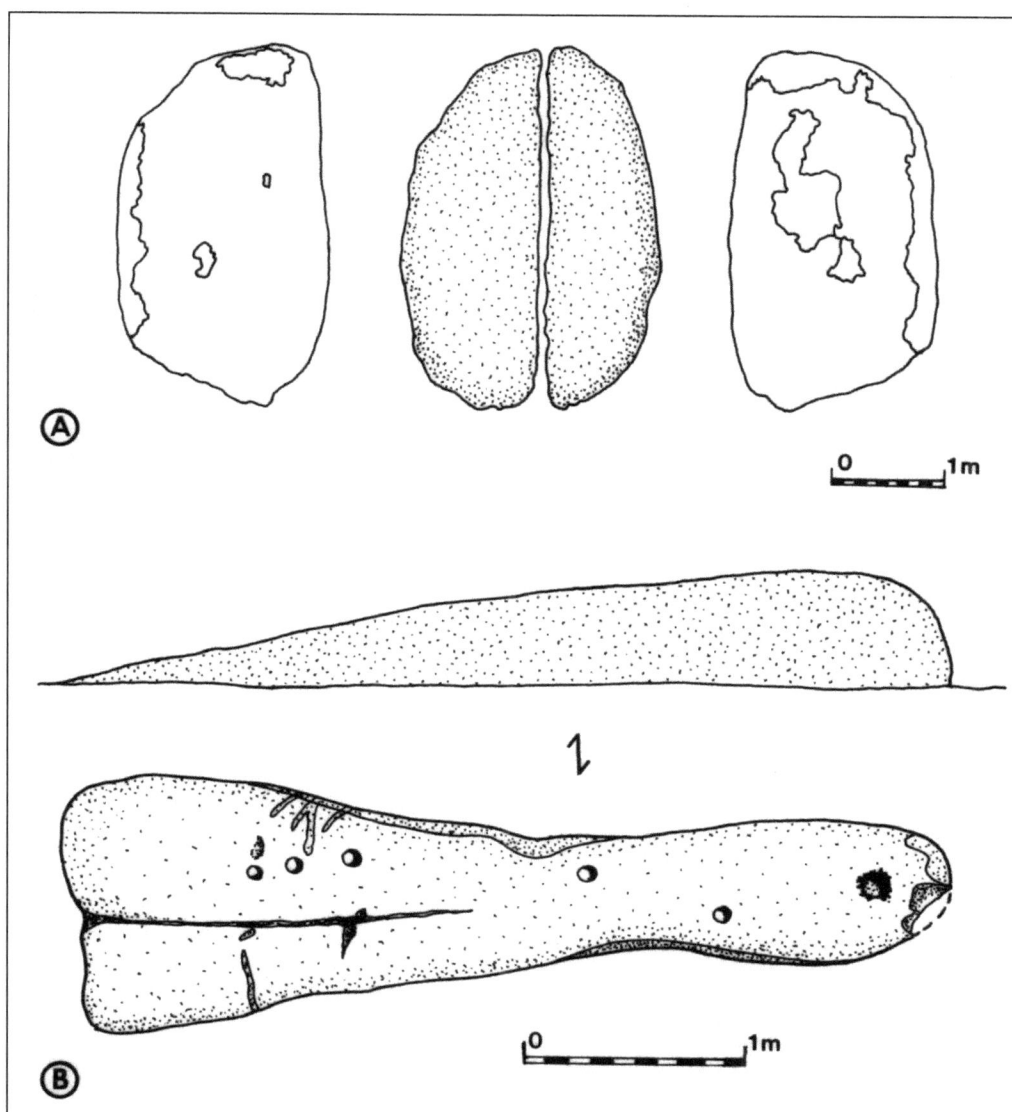

Fig. 6.3. Opportunisme et choix des blocs: a) ovoïde de la pierre; b) bloc anthropomorphe à cupules des Prés Bas

DEBITAGES, TRACTION ET LEVAGE

La majorité des grands menhirs provient des affleurements oblongs qui présentent des fentes de décompression, généralement parallèles à la surface du sol. Ce choix des paléo-formes stéréotypées, déterminé par la monumentalité, les silhouettes recherchées et les modalités techniques d'arrachage, engendre des aboutissants architectoniques très comparables où les stigmates d'extraction se reconnaissent facilement. Deux modes de préparation ont été reconnus (Fig. 6.4):

1) l'encochage et le forçage sur fente

2) l'éclatement par saignée ou rainure, complété par encochage et forçage.

Ces deux procédures techniques utilisent invariablement les failles de décompression des granites. Pourtant, quelques rares grands menhirs ont également été dégagés

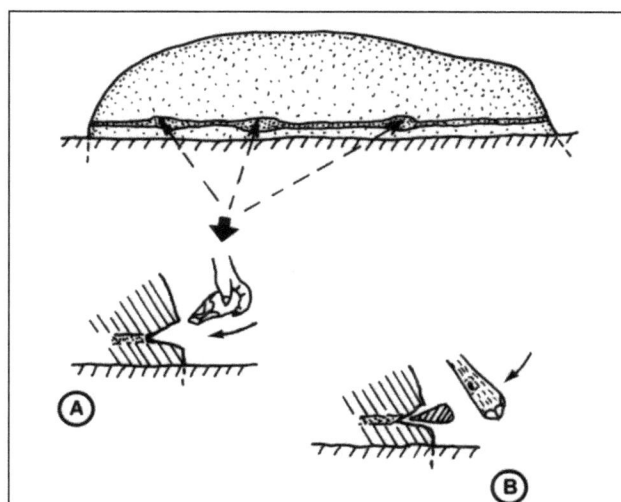

Fig. 6.4. Principe de débitage de la roche
(d'après observations)

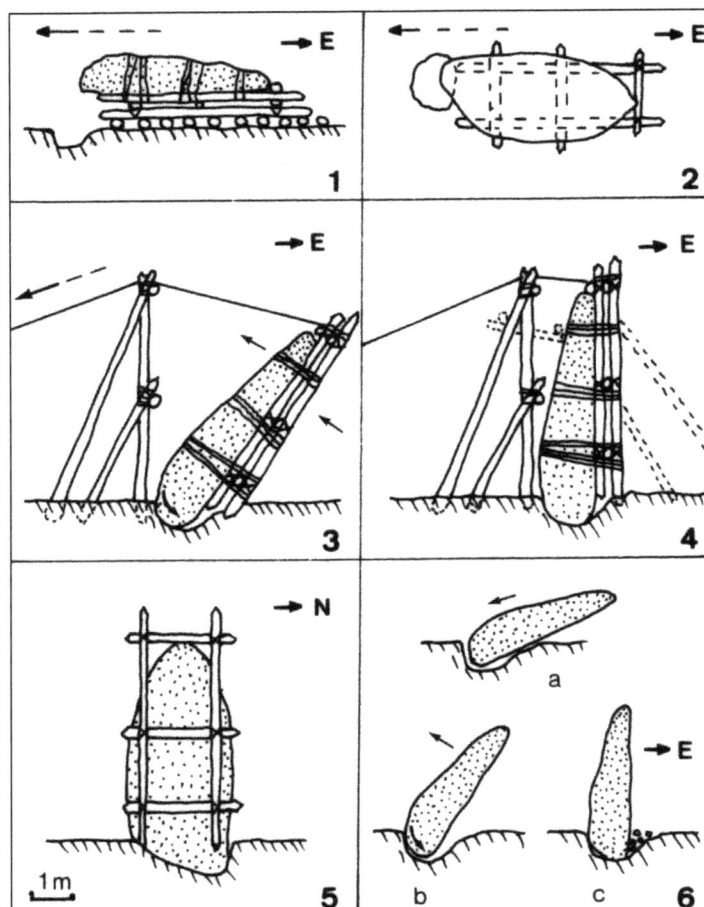

Fig. 6.5. Modélisation d'un levage d'après les données de fouilles

dans d'épaisses plaques de grès erratiques, au moyen de la technique de rainurage et forçage (Benéteau 1999, 193).

Le transport de ces énormes monolithes, des carrières à leurs lieux d'érection, parfois sur des pentes à 10% d'inclinaison et sur plusieurs centaines de mètres (jusqu'à 1000 mètres), ne pouvait être réalisé sans une armature de bois conçue pour chacun d'entre-deux.

Le premier "choc technique" consistait à riper le bloc, hors de la zone d'extraction, sans doute sur un caillebotis ou un plancher, aboutissant à un traîneau. Ce traîneau permettra non seulement un cheminement sur rouleaux, très facilité, mais il va aussi participer à l'armature du levage et à l'immobilisation du menhir lors de l'érection. Toutes les traces de cette dernière phase ont été observées lors de la fouille de l'alignement G2 du Bois de Fourgon à Avrillé (Benéteau 1999, 223-4). Après fendage, il est évident que la face du bloc arraché, côté issu de la roche-mère, dite "lit de carrière", se trouve en situation de glissement favorable, permettant une fixation aisée sur le traîneau (la face plate vers le sol). Les mégalitheurs ont parfaitement compris que le transport de telles masses s'effectue nettement plus facilement, lorsque celles-ci sont posées sur leur face plane, plutôt que sur la face irrégulière ou arrondie, bien plus difficile à arrimer.

Pour un bloc de 25 tonnes, deux troncs d'arbres d'une trentaine de centimètres de diamètre, reliés entre-eux par trois entretoises d'une vingtaine de centimètres de diamètre, suffisent à réaliser un traîneau rudimentaire mais efficace, sur lequel, une fois attaché, celui-ci formera un ensemble solidaire favorisant le guidage du chargement jusqu'au lieu de levage. L'utilisation de cette technique génère une notable réduction des efforts nécessaires pour vaincre l'inertie des masses à déplacer (à plus forte raison sur des pentes accidentées). De plus, lors de la poussée, l'implantation volontaire par "butée", dans le sol, des extrémités du traîneau, participe mécaniquement au levage par un effet en "béquille", mis en mouvement par la "traction en chèvre", démultipliée par un puissant chevalet dont les traces d'encrage dans le sol ont été, elles aussi, constatées à la fouille (le diamètre des poteaux latéraux de ce chevalet avoisinait les 50 centimètres).

Cette armature de bois parait avoir également servi à stabiliser le menhir en position verticale, par un échafaudage en "berceau" (Fig. 6.5). Au levage, le monolithe, qui présente donc sa face plate côté sol, est soulevé en direction de l'ouest (coucher du soleil), orientant automatiquement cette face plate à l'est (lever du soleil), et démontre ainsi, que tous les levages des menhirs de cette région ont été réalisés de cette façon, puisqu'ils montrent tous leurs faces plates à l'est.

L'implantation des bases des menhirs dans les fosses, soutient cette interprétation. En effet, lors de sa montée (face plate au sol: traîneau), le bloc bascule dans un premier temps dans la fosse, dont la bordure est écrasée par le poids ; sa base talonne alors la paroi opposée, ce qui enclenche un effet de contrebutée. Le menhir, en se redressant, prend appui sur la paroi ouest de la fosse et projette sa face plate vers l'est, alors que par ripage, dû à la traction en flèche du bloc, la base de la face bombée est plaquée contre la paroi ouest (Fig. 6.6). Il en résulte qu'il n'existe plus d'espace pour caler le menhir dans sa fosse, que du côté est.

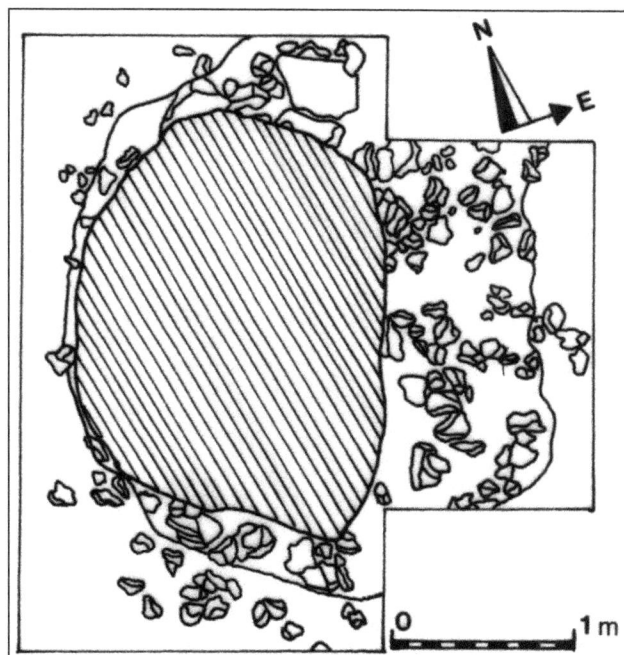

Fig. 6.6. Exemple d'implantation d'un menhir dans sa fosse (menhir du Plessis, Le Bernard, Vendée)

Toutes les bases des grands menhirs fouillés ont été bloquées de cette façon et ces constatations amènent une réflexion primordiale. Les techniques d'implantation ayant été réalisées de façon rigoureusement identiques, on peut donc raisonnablement supposer que ces architectures de pierres dressées ont été construites au cours d'un intervalle de temps relativement court, une génération ou deux, sinon la technique aurait évolué et les modifications architectoniques engendrées seraient perceptibles à la fouille (ce qui n'est pas le cas).

ESTHETIQUE ET SEXUALISATION

Le surfaçage des monolithes, dans leur ensemble, reste sommaire, même si on observe pour certains d'entre-eux

un martelage plus soigné. L'épannelage des traces d'extraction et l'écrasement des parties anguleuses n'affectent que certaines zones des blocs, mettant en évidence une sexualisation de ces pierres dressées.

En réalité, les bouchardages réguliers de la roche sont visibles principalement sur les faîtes et les parties basses, ce qui accentue, pour certains menhirs, la "céphalisation" de leur sommet, tout en maintenant une base brute ou du moins assez large, par rapport à l'ensemble de la silhouette. Pour d'autres, au contraire, la base sera "amincie" (on aura sans doute amplifié la forme préchoisie), laissant la "tête" sans finition soignée. Quelques-uns de ces grands menhirs paraissent cependant avoir bénéficiés d'une régularisation complète de leur surface (menhir 2 du Plessis, commune du Bernard, menhir de La Chenillée, commune de St Sornin) (Fig. 6.7).

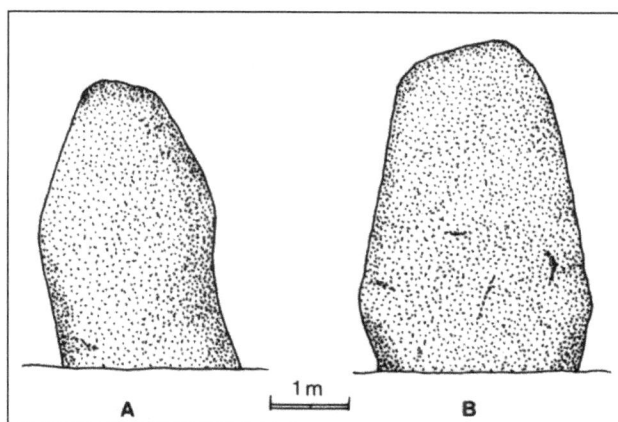

Fig. 6.7. Sexualisation des menhirs

Ce rétrécissement des bases évoque fortement la silhouette d'une femme dont les hanches sont, pour des raisons naturelles évidentes, plus développées que celles des hommes, et de fait, créé une certaine gracilité dans l'anthropomorphisme ainsi codifié, de ces monolithes. A contrario de ces "hanchements", les blocs masculinisés, plus massifs, moins travaillés, montrent par contre, des sommets plus "pointus", fréquemment ogivaux: La Boilière, Avrillé, le Bois de Fourgon, Avrillé (Benéteau 2006, 576).

Quelques rares gros menhirs, plus bruts, semblent aussi s'écarter de ces "canons" originaux, d'autres présentent quelques gravures grossières réalisées par rainurage profond de la roche (cou du menhir anthropomorphe de l'alignement de La Pierre qui Vire, à Longeville).

CONCLUSION

La place des grands menhirs anthropomorphes sexualisés dans ces alignements semble bien participer à une dimension symbolique prégnante de cette société de levageurs préhistoriques. Leur implantation monumentale dans l'espace environnemental (sur les élévations de terrain) et leur assemblage en "groupes d'alignements", évoquent une sorte d'architectonie majestueuse d'un ordre social élaboré. En corollaire, le fait que le "corps modèle" de "l'homme géométrique" ait sans doute servi d'étalon dimensionnel dans l'architectonique et l'architecture de ces constructions, a probablement influencé la vision de ce que nous percevons comme pouvant être un "style" architectural mégalithique. Pourtant, rien n'est moins sûr.

Par contre, la nouvelle problématique à développer, en matière de mégalithisme, reste celle des carrières. Il est, en effet, de plus en plus évident, malgré les trop rares fouilles réalisées sur de tels sites (Lecerf 1999; Benéteau 1999; Mens 2002), que l'ampleur des travaux engendrés dans ces carrières soit sous-estimée, tant par les diverses (et parfois complexes) modalités d'extraction, que par les ouvrages de bois, liés au transport et, très certainement dans certains cas, aux terrassements importants permettant de décaisser les affleurements ou d'évacuer les blocs hors des gisements.

Les surfaces ainsi occupées, paraissent de plus en plus étendues au fur et à mesure que l'on s'approprie une expérience en ce domaine. De même, l'outillage des mégalitheurs doit faire l'objet d'un protocole spécifique d'identification et de description. Bien que plusieurs de ces outils aient été rencontrés et reconnus lors de rares fouilles de carrières (Lecerf 1999), il existe un grand vide typologique en ce domaine. Ces outils révèlent une gamme très variée, tant par la diversité des formes et des volumes que par la nature des matériaux utilisés et des roches à exploiter. La fouille complète d'une carrière devra être entreprise, si possible en relation avec un/des monument(s) mégalithique(s) proche(s).

Plus difficile, car il faudra s'attaquer à des tabous en préhistoire, va être la recherche d'une relation possible entre de petites compositions de cupules et les blocs mégalithiques, comme pouvant constituer un mode de "marquage", effectué en amont des extractions, peut-être

même lors de la délimitation des bancs rocheux à exploiter.

Ces "nouvelles problématiques" (et il en reste d'autres à proposer), doivent favoriser l'exploration de voies innovantes en matière de recherche sur le mégalithisme, réinterroger des constructions déjà étudiées et définir d'autres approches sur des architectures modestes d'apparence, mais pouvant devenir complexes à la fouille.

Bibliographie

BENETEAU, G. (1998) Les grands menhirs du Talmondais. In Joussaume, R., dir, *Les premiers paysans du golfe: le Néolithique dans le Marais Poitevin*. Chauray: Editions Patrimoine et Médias, 96-99.

BENETEAU, G. (1999) *Les alignements de menhirs du sud de la Vendée*. La Roche-sur-Yon : Editions Anthropologica.

BENETEAU, G. (2006) Les alignements de menhirs anthropomorphes du sud de la Vendée, architectonique, iconographie et art pariétal. In Joussaume, R., Laporte, L., & Scarre, C., dir, *Origine et développement du mégalithisme de l'ouest de l'Europe, actes du colloque international de Bougon, 26-30 octobre 2002*. Bougon: Musée des Tumulus de Bougon, 569-577.

JOUSSAUME, R. (2003) *Les Charpentiers de la Pierre, les mégalithismes dans le monde*. Paris: La Maison des Roches.

LECERF, Y. (1999) *Monteneuf. Les Pierres Droites. Réflexions autour des menhirs*. Rennes: Documents Archéologiques de l'Ouest.

MENS, E. (2002) *L'affleurement partagé, gestion du matériau mégalithique et chronologie de ses représentations gravées dans le Néolithique moyen armoricain*. Thèse de doctorat en archéologie, Université de Nantes.

TECHNOLOGIE DES MEGALITHES DANS L'OUEST DE LA FRANCE: LA CARRIERE DU ROCHER MOUTON A BESNE (LOIRE-ATLANTIQUE, FRANCE)

Emmanuel MENS

Collaborateur UMR 6566 du CNRS, 19 rue du Vieux Moulin, 44740 Batz-sur-Mer, France

Abstract: By means of typology, megalithic blocks can be mentally repositioned in the upper and lower levels of the outcrop from which they came, in the same way that flint flakes can be repositioned in a core. As a result, a mental refitting of the initial outcrop is possible and this justifies the use of the term "core-outcrop" for the original rock outcrop that is reconstructed in this way. A technological study of the working of these stones is proposed, based mainly on the relocation of each megalithic block in its original position within the sequence of successive stages in the working of the outcrop. This method opens up a new field of research on the deployment of megalithic raw material during the Neolithic period, and initial results are presented from the excavation of the megalithic quarry of Rocher Mouton at Besné (Loire-Atlantique, France).
Key-words: megalithic technology, mental refitting, megalithic quarries

Résumé: A l'aide d'une typologie, les blocs mégalithiques peuvent être repositionnés dans les étages supérieurs ou inférieurs du rocher à la manière d'éclats de silex détachés de leur nucléus. Un remontage mental théorique de l'affleurement initial est alors possible et justifie le terme "d'affleurement nucléus" pour désigner le rocher ainsi reconstitué. Une étude technologique du débitage de ces rochers est proposée en s'appuyant principalement sur la restitution de l'origine de chaque bloc mégalithique dans le déroulement du processus de partage de l'affleurement. Cette méthode ouvre alors un nouveau champ de recherche sur la gestion du matériau mégalithique au Néolithique, dont les premiers résultats sont présentés avec la fouille de la carrière mégalithique du Rocher Mouton à Besné (Loire-Atlantique, France).
Mots clés: technologie des mégalithes, remontage mental, carrières mégalithiques

Une approche originale des premières architectures en pierres dans l'Ouest de la France est proposée à l'aide d'une lecture technologique. Ce type d'étude, dont on connaît le succès sur les industries en silex, permet de renouveler en profondeur les problématiques de recherche sur le mégalithisme.

L'étude technologique s'attache à reconnaître les processus d'acquisition, de transformation et d'utilisation des blocs mégalithiques provenant du partage d'affleurements de nature granitique. Suivre la gestion des menhirs, des orthostats, et des tables de couverture issus du débitage d'un même affleurement est un angle de lecture qui n'a jamais été exploré jusqu'à présent. La nouveauté de la méthode consiste à resituer l'origine de chaque bloc mégalithique dans le déroulement du processus de partage de l'affleurement grâce au remontage mental. Cette méthode s'appuie sur la forme spécifique des blocs mégalithiques provenant des rochers à érosion sphéroïdale comme le granite. L'identification conjuguée des plans de débits et des anciennes faces d'affleurement permet de reconnaître l'emplacement du bloc dans le rocher avant son extraction. Les blocs mégalithiques peuvent ainsi être repositionnés dans les étages supérieurs ou inférieurs du rocher à la manière d'éclats de silex détachés de leur nucléus.

A l'aide de cette lecture technologique, l'objectif est d'analyser toutes les différentes phases qui ont contribué à la construction des premières architectures en pierre. Reconstituer ainsi la chaîne opératoire des blocs mégalithiques depuis leur extraction jusqu'à leur utilisation finale permet de recueillir des informations fiables sur le contexte économique et social ayant favorisé l'éclosion et le développement de l'architecture monumentale. Le remontage des affleurements ayant servi à la construction des mégalithes ouvre un nouveau champ de recherche sur la gestion du débitage du matériau de construction au Néolithique.

Parmi les objectifs prioritaires de ce nouveau champ de recherche, la question des techniques d'acquisition du matériau de construction tient une place centrale. Si des travaux et des expériences d'extraction réussie ont été menés à bien sur les roches calcaires (Mohen & Scarre 2002), en revanche à ce jour, sur les roches plus dures du Massif armoricain, seules des fouilles de carrières mégalithiques ont été réalisées, notamment dans le Sud de la Vendée (Bénéteau, 2000). Deux essais d'extraction ont néanmoins été tentés par le Groupe Archéologique de Saint-Nazaire. La dernière tentative a été menée à l'aide des observations recueillis au cours de la fouille de la carrière mégalithique du Rocher Mouton à Besné (Loire-Atlantique, France). Cette expérience a permis de reproduire les éclats thermiques récoltés en grand nombre au moment de l'exploration du site, conduisant à privilégier l'utilisation du feu par chauffage de la roche pour le détourage et l'arrachement des dalles mégalithiques.

LES SOURCES D'APPROVISIONNEMENT DANS L'OUEST DE LA FRANCE: EXTRAIRE PLUTOT QUE RECUPERER DES BLOCS ERRATIQUES

Sur les substrats granitiques de l'Ouest de la France, les chercheurs ont souvent insisté sur le transport des blocs, mais beaucoup moins sur leur extraction. Sans doute en raison d'une idée très répandue selon laquelle les premières communautés agro-pastorales auraient fait preuve d'une stratégie totalement opportuniste dans l'acquisition des matériaux: l'approvisionnement se satisfaisant des blocs déjà isolés, en partie ou en totalité, de la roche massive par des fissures géologiques (Bessac 1998). Le caractère très souvent brut de débitage des dalles qui ont conservé leur face naturelle en forme de boule et de chaos granitique jouant logiquement en faveur de cette thèse.

Aussi, avant même d'évoquer la question des carrières mégalithiques, il est important de souligner qu'une étude portant sur un corpus de 2000 blocs observés entre l'estuaire de la Loire et la région de Carnac (Morbihan), a montré que l'immense majorité des dalles est le fruit d'une intervention significative de l'homme sur la matière, et non la récupération de pierres naturellement détachées du substrat (Mens 2002). Le propos n'est pas de nier, qu'effectivement dans certains cas les constructeurs ont bien utilisé des pierres déjà entièrement déchaussées par le jeu de l'érosion, notamment sur l'estran des régions littorales, mais de préciser que ces cas sont minoritaires. De plus, il serait faux de laisser entendre que les constructeurs des zones littorales ont systématiquement fait preuve d'opportunisme. L'exemple de l'alignement de menhirs du Douet sur l'île d'Hoedic (Morbihan: travaux en cours sous la direction de Jean-Marc Large) et de la tombe à couloir de Dissignac (Saint-Nazaire, Loire-Atlantique) montre que la question du choix du matériau est beaucoup plus complexe que les simples problématiques de distance parcourue ou de résistance mécanique. A Dissignac, une partie des blocs mégalithiques vient de l'estran situé actuellement à 4 km, soit environ 4,5 km à l'époque néolithique. Au Douet, les huit blocs dressées ont une origine terrestre alors que l'estran actuel est à 10 mètres, soit environ 500 mètres au Néolithique (Large 2004). Au Douet, les constructeurs choisissent l'extraction terrestre alors que l'estran est tout proche. La provenance terrestre des blocs dans un contexte aussi proche de l'estran souligne bien le caractère non opportuniste de l'accès aux ressources. Manifestement, d'autres critères ont influencé les constructeurs dans le choix de l'affleurement, parmi lesquels ceux de la forme, de la couleur, ou de la texture de la pierre ont pris une part non négligeable.

Entre l'estuaire de la Loire et le Golfe du Morbihan, si l'extraction du matériau de construction est la norme pour alimenter la grande densité de monuments mégalithiques, en revanche, la récupération de dalles déchaussées naturellement par le jeu de l'érosion est exceptionnelle. Ce choix de l'extraction traduit une "pression" sur le substrat exercée par une tradition d'architecture en pierres longue de trois millénaires. Dans la région de Carnac (Morbihan), les toutes premières extractions néolithiques ont pu effectivement profiter des blocs détachés naturellement du substrat. Mais dans un contexte de relief rocheux résiduel, les extractions néolithiques secondaires, faute de blocs erratiques en quantité suffisante pour répondre à l'énorme demande, ont rapidement commencé à débiter les affleurements au-dessus du sol. Puis, une fois les

affleurements au-dessus du sol épuisés, les extractions ont alors attaqué le substrat en profondeur. Une chronologie technique du débitage en est déduite où les extractions primaires sont caractérisées par une fréquence importante des blocs issus de l'étage supérieur de l'affleurement, alors que les extractions postérieures sont quant à elles, réduites aux blocs de l'étage inférieur, en particulier les blocs sans face d'affleurement qui sont le fruit d'un décaissement de plus en plus profond (Mens 2007).

L'étude des méthodes de débitage dans les alignements de Carnac a montré une gestion raisonnée des affleurements (Mens 2002). Cette gestion est en contradiction totale avec la conception opportuniste souvent proposée lorsque l'on évoque la relation entre les premières sociétés agro-pastorales porteuses d'une architecture monumentale et leur substrat rocheux. Il apparaît que la méthode de débitage à Carnac favorise principalement le plan de débit horizontal. Ce choix technique est très certainement lié à la nécessité de conjuguer des besoins importants en pierres de grande taille (beaucoup de menhirs font plus de trois mètres de haut), avec des affleurements en nombre limité dont la hauteur dépassait rarement deux mètres.

Le débitage de ces affleurements n'a pas manqué de laisser des traces sur les blocs qui en ont été extraits. Aussi, malgré la fracturation non conchoïdale du matériau granitique souvent employé sur la façade ouest de la France, différents niveaux de lectures permettent d'identifier les stigmates d'une extraction préhistorique.

Le premier niveau de lecture concerne la présence d'une ancienne face d'affleurement opposée à une face d'arrachement, car sur les 2000 blocs étudiés, 60% d'entre eux proviennent des étages supérieurs de l'affleurement. Le deuxième niveau de lecture concerne la reconnaissance du caractère anthropique de la face d'arrachement. Dans ce cas, une rupture anguleuse relie la face d'arrachement à la face d'affleurement. Cette rupture de forme est associée à des enlèvements de matière ou à des encoches de débitage. Ces ruptures de forme à la jonction de l'ancienne face d'affleurement et de la face d'arrachement plaident pour une intervention anthropique. Parfois, le plan de débit de la face d'arrachement forme des petites "marches" très caractéristiques. Enfin, toujours à la jonction entre la face d'arrachement et la face d'affleurement, on observe des taches de rubéfaction qui peuvent être les restes d'un choc thermique effectué au moment de l'extraction. Lorsque tous ces éléments sont réunis sur un même bloc et que ce dernier entre dans l'une des catégories typo-technologiques du remontage mental de l'affleurement, la présence d'un produit de débitage préhistorique doit être sérieusement envisagé.

LA THEORIE DU REMONTAGE MENTAL DE L'AFFLEUREMENT

Un bloc mégalithique en provenance d'un affleurement à altération sphéroïdale présente deux grandes faces princi-

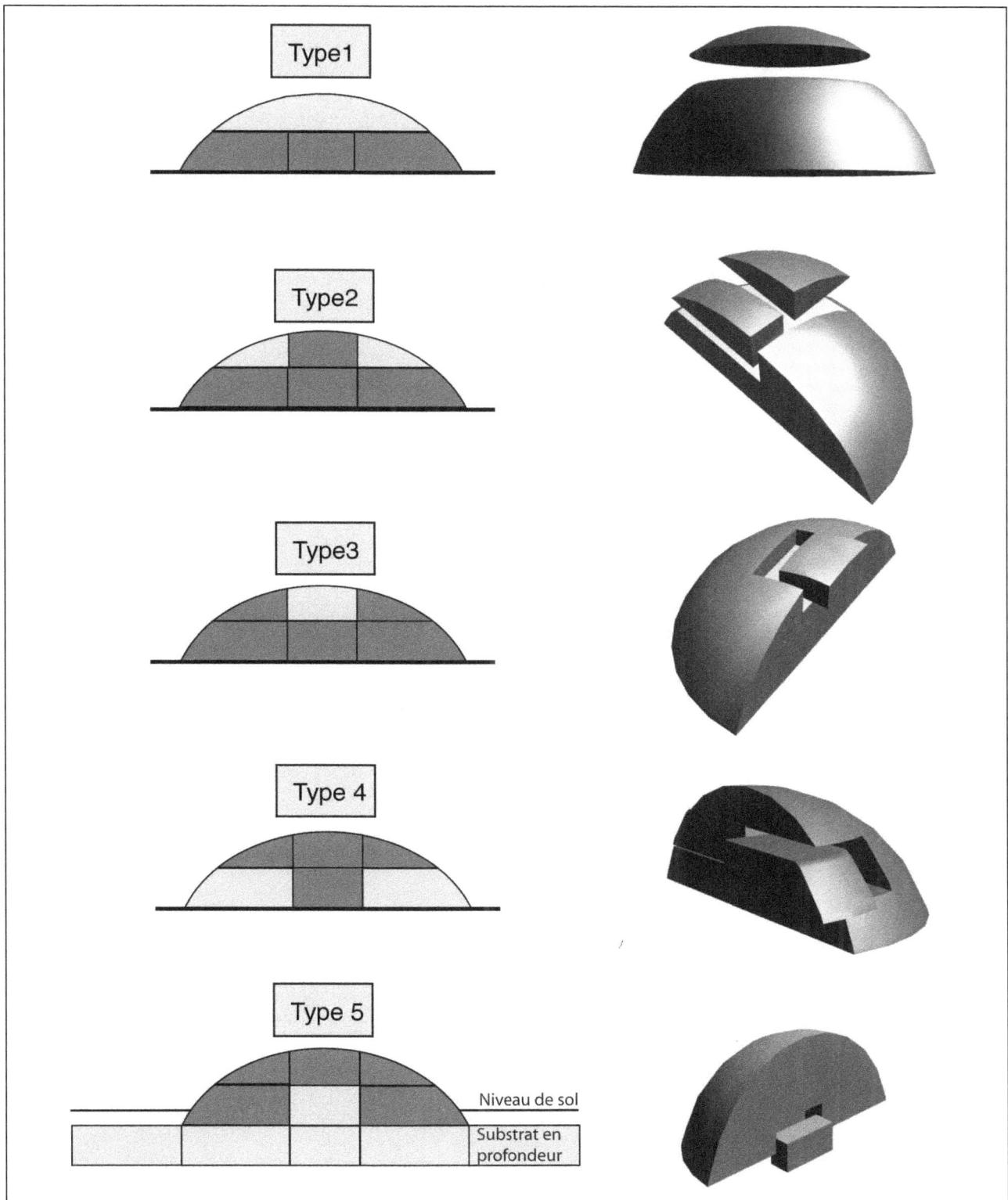

Fig. 7.1. Typologie du remontage mental de l'affleurement nucleus

pales: une face d'arrachement presque plane initialement engagée dans le rocher, et une ancienne face d'affleurement de forme convexe, correspondant à la partie du rocher initialement exposée à l'air libre (Sellier 1991, 1995).

L'identification conjuguée des diaclases utilisées comme plans de débits et de l'ancienne face d'affleurement constitue la base du système de description. L'ancienne face d'affleurement à la forme convexe est l'élément principal d'information permettant de reconnaître l'emplacement du bloc dans le rocher naturel avant son extraction, servant en quelque sorte de "néocortex". A l'aide du système de fractionnement naturel du granite, par le jeu des diaclases employées comme plans de débits qui s'articulent autour des faces d'affleurements, il est alors possible d'établir une déclinaison théorique et typo-technologique des possibilités d'extraction de l'affleurement (Mens 2005).

Lorsque l'affleurement initial est suffisamment haut, son partage permet de distinguer cinq grands types (figure 7.1): les types 1,2 et 3 ont une face d'affleurement comme face principale et proviennent des étages supérieurs du rocher; les types 4 et 5 ont des faces principales essenti-ellement formées de faces d'arrachement et sont issus des étages inférieurs de l'affleurement.

L'intérêt de la typologie est de décrire les blocs en fonction de leur appartenance aux différents étages d'extraction, permettant de raisonner en terme d'ordre d'apparition. Ainsi, au fil de la chaîne opératoire, les blocs en provenance des étages supérieurs de type 1 et 2 sont-ils systématiquement enlevés avant ceux de type 4 et 5 (Fig. 7.1). On peut alors proposer un modèle théorique

de remontage mental de l'affleurement inspiré des méthodes développées par les lithiciens, lorsque les raccords réels entre les éclats de silex ne sont pas envisageables (Pélegrin 1995). Cette méthode empruntée aux spécialistes des roches siliceuses justifie le terme "d'affleurement nucléus" pour désigner le rocher initial ainsi reconstitué.

Si les méthodes de débitage ont pu être décrites avec précision grâce au remontage mental de l'affleurement (Mens 2002), en revanche, les techniques de débitages restent en grande partie à explorer. Afin de répondre à cette question, la fouille de la carrière mégalithique du Rocher Mouton (Loire-Atlantique) a débutée en 2005 avec le concours du Groupe Archéologique de Saint-Nazaire. Les résultats présentés ici sont préliminaires à la publication exhaustive des fouilles 2005-2006.

LE SITE DU ROCHER MOUTON A BESNE (LOIRE-ATLANTIQUE, FRANCE)

Le Rocher Mouton est situé au Nord-est du Marais de Brière, vaste dépression marécageuse limitée au Sud-

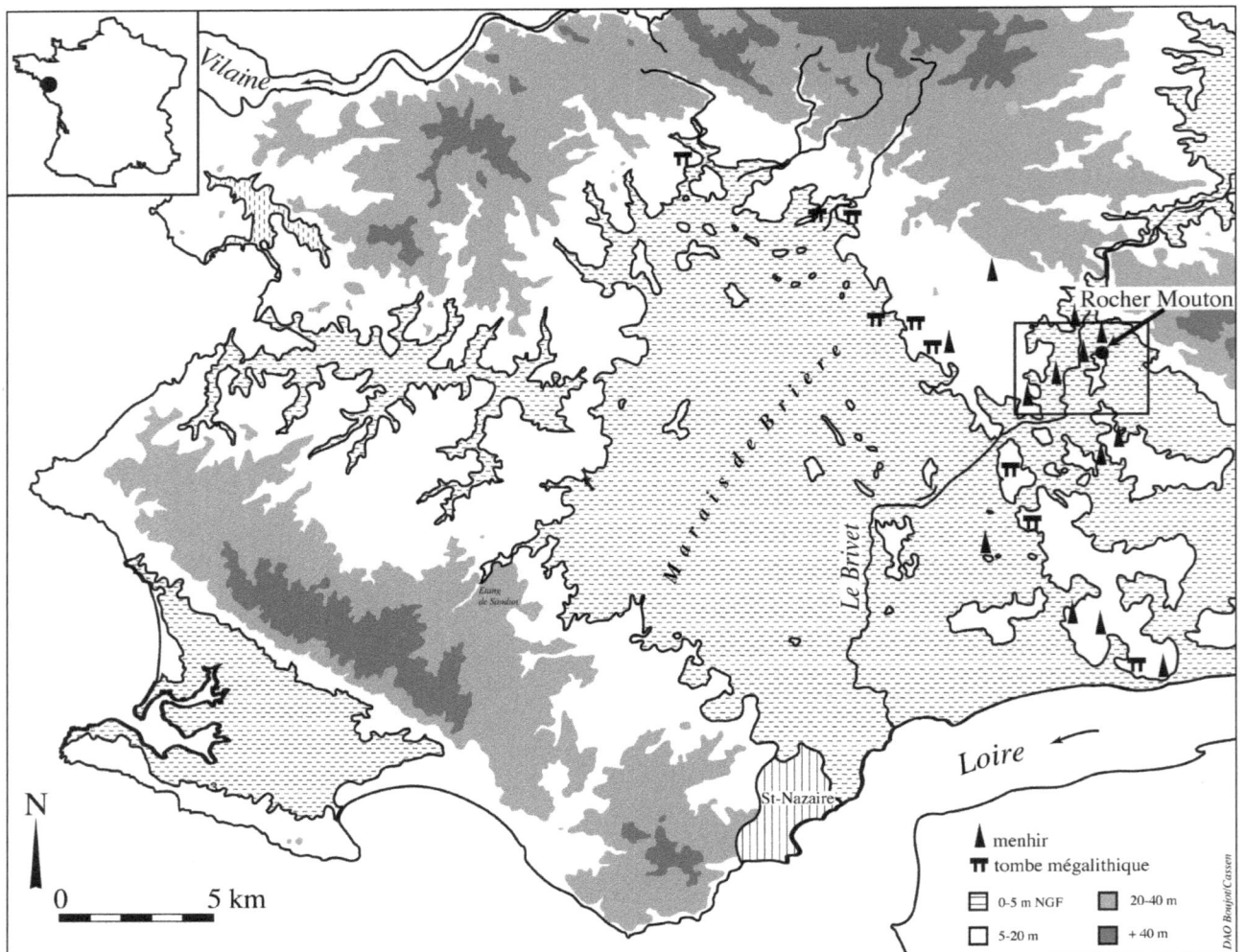

Fig. 7.2. Cartographie des mégalithes autour du Rocher Mouton (non exhaustif entre Loire et Vilaine)

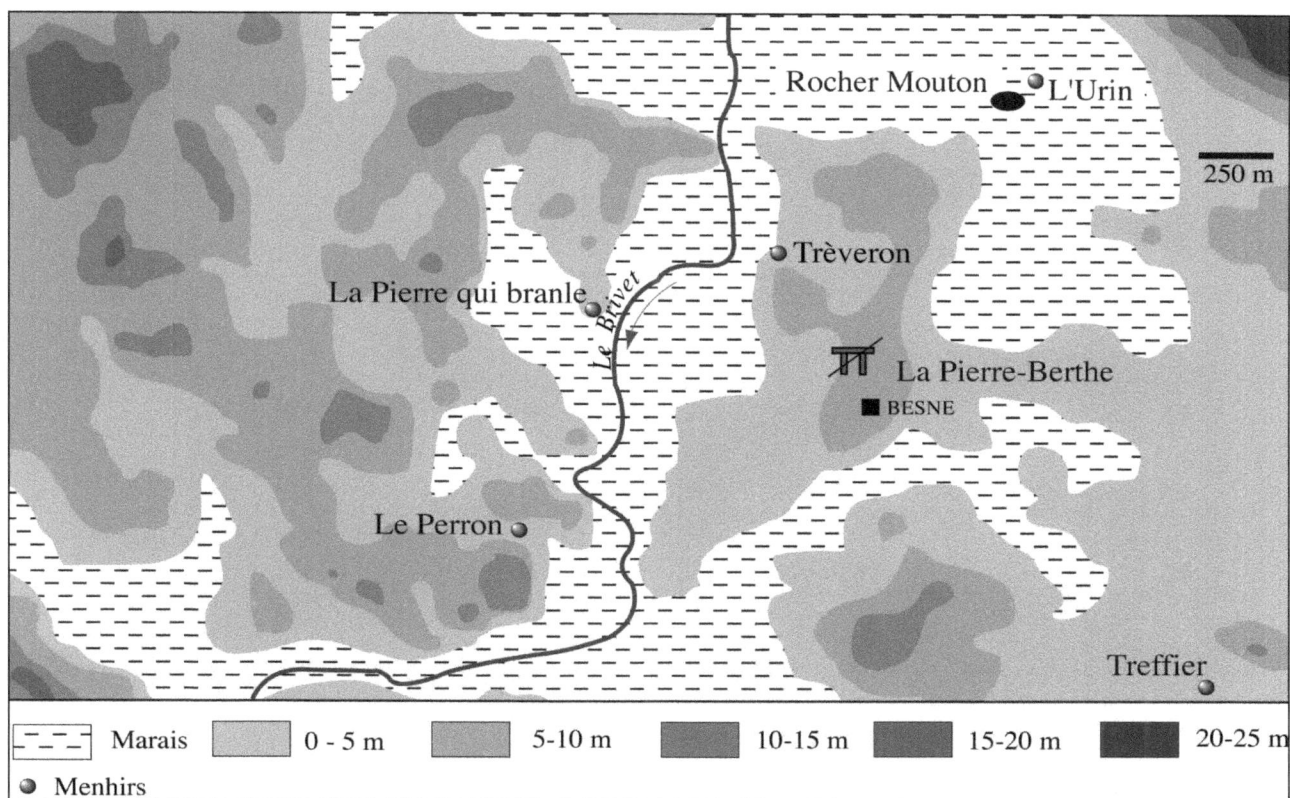

Fig. 7.3. Carte de localisation du Rocher Mouton

ouest par le massif granitique guérandais, au Sud par la Loire et au Nord-est par le sillon de Bretagne (Fig. 7.2). Le Rocher Mouton intègre la zone médiane du Bassin du Brivet, petit fleuve d'une quarantaine de kilomètres longeant le site à 500 mètres à l'Ouest et qui draine ensuite le marais de Brière (Figs. 7.2 et 7.3). Au Néolithique moyen, la dépression briéronne est pénétrée par l'eau de mer et ressemble à un immense estuaire où se déposent des vases bleues. Une phase de régression vers 4500 BP voit l'exondation des vases bleues et l'implantation d'une forêt. La mise en place de la tourbe brune se fait vers 4200 BP avec l'apparition d'un cordon vaseux qui isole la dépression de l'estuaire (Prigent 1978; Visset 1982, 1990). Si, aujourd'hui, le Rocher Mouton est situé à l'intérieur des terres, en revanche au Néolithique moyen, par sa proximité avec le Brivet, il est en contact direct avec l'immense estuaire ligérien formé alors par la dépression briéronne.

Le substrat du Bassin du Brivet est composé pour partie d'un granite d'anatexie à biotite, également appelé "granite de Prinquiau". A l'image du Rocher Mouton, le secteur géographique du Bassin du Brivet situé au pied du sillon de Bretagne voit de nombreuses petites buttes émergeant du marais dont les sommets sont ponctués d'affleurements en granite de Prinquiau (principalement les communes de Besné et Pontchâteau).

Le site du Rocher Mouton s'inscrit au centre d'une région particulièrement dense en mégalithes, où la concentration

de menhirs est l'une des plus importantes entre Loire et Vilaine (L'Helgouac'h 2001). On notera au passage dans ce secteur géographique une distribution spatiale fort différente entre les pierres dressées et les tombes à couloir (Fig. 7.2). Si ces dernières apparaissent uniquement sur les bords du marais de Brière, en revanche, les pierres dressées ont une répartition plus diffuse, avec une concentration au Nord dans les zones humides afférentes aux marais du Brivet. Les études de Daniel Prigent et Lionel Visset (1977) effectuées sur les menhirs de la Pierre Blanche (Trignac) et de Hêlé (Donges) ont montré que les blocs ont été dressés lors d'un épisode régressif de la transgression flandrienne, avant la formation de la tourbe qui s'installe vers 4200 BP.

Le périmètre de trois kilomètres autour du Rocher Mouton compte de nombreux menhirs. Parmi ceux encore visibles, citons le menhir de Treffier (Besné), de Tréveron (Besné), de la Jourdanais (Pontchâteau), de l'Urin (Pontchâteau), de Hêlé (Donges) et du Perron (Pontchâteau). Parmi ces pierres dressées, on soulignera la présence du menhir de l'Urin situé dans le marais à une centaine de mètres au Nord-est du Rocher Mouton (Fig. 7.3).

Le Rocher Mouton est une petite butte qui émerge du marais à 2 mètres NGF, avec à son sommet six affleurements en "granite de Prinquiau". Ce granite présente des grains hétérométriques souvent grossiers où les cristaux de feldspath ont de remarquables formes quadrangulaires ainsi qu'une taille pouvant aller jusqu'à 10 cm de long.

Fig. 7.4. Le bloc déconnecté du rocher nord (vue de l'Est)

Les affleurements du Rocher Mouton montrent un réseau de fissuration particulièrement resserré rarement observé sur les autres rochers du secteur. Sur cette butte, la longueur des affleurements va de trois à onze mètres et la hauteur maximum est d'un peu moins de deux mètres. Deux affleurement ont été retenu, l'un au Nord, mesure onze mètres de long et huit mètres de large, l'autre au Sud fait 6,5 mètres de long, 6 mètres de large et 1,95 mètres de haut (Fig. 7.6).

Le choix du site

Dans la perspective d'explorer en détail les techniques d'extraction de mégalithes, le site s'est imposé pour plusieurs raisons. D'une part, l'évidence de la concentration de pierres dressées autour du Rocher Mouton a constitué un élément non négligeable dans le choix du site à tester (Figs. 7.2 et 7.3). D'autre part, des observations faites avant la fouille, à fois géomorphologiques et technologiques, vont toutes dans le sens de l'existence d'une carrière ancienne. Par exemple, la présence d'un bloc de taille mégalithique déconnecté du substrat au sommet du rocher nord est un élément qui a été pris en compte (Figs. 7.4 et 7.6). Ce bloc mesure 3,55 mètres de long, 0,90 mètres de large et 0,85 cm d'épaisseur avec un espace entre la roche mère qui atteint parfois 40 cm. La pierre a été détachée puis déplacée de quelques dizaines de centimètres comme l'atteste ses

diaclases légèrement obliques par rapport à celles de l'affleurement. Le bloc est parcouru de plans de débits sauf à son sommet constitué d'une face d'affleurement. La pierre est coupée de chaque côté par des plans de débits verticaux, ce qui correspond au schéma d'extraction de type 3 (Fig. 7.1). En l'absence de toutes traces d'outils métalliques sur ce bloc, la question de la présence d'une carrière mégalithique a pu légitimement être posée.

D'autres indices découverts avant la fouille lors d'un simple nettoyage des mousses sur les autres affleurements du site plaident également pour une extraction ancienne. Par exemple, aucune trace de mortaise circulaire de type barre à mine n'a été détectée le long des fronts de taille verticaux, excluant ainsi une exploitation moderne. Toujours dans le sens d'une extraction ancienne, certains fronts de taille montrent des traces d'érosion que l'on retrouve souvent sur les faces d'arrachement des menhirs. Deux parois verticales montrent ainsi des réseaux polygonaux qui caractérisent des formes anciennes d'érosion postmégalithiques (Sellier 1997). Au moment d'ouvrir les sondages, la priorité accordée au pied des parois montrant ces réseaux polygonaux s'est avérée payante. Dans le même registre, la présence de cuvettes d'érosion spatialement associées à des fronts de taille est également un signe d'ancienneté prouvée ensuite par la fouille.

Fig. 7.5. Percuteur en grès engagé dans une fissure naturelle élargie du Rocher nord

D'autres indices plus directement liés à l'extraction préhistorique sont directement visibles sur l'affleurement après un simple nettoyage. Par exemple, la présence de gorges de débitage dont la section adoucie révèle l'utilisation d'une percussion lithique, ou la découverte de percuteurs en grès ou en quartz encore engagés dans les fissures naturelles de la roche (Fig. 7.5), sont autant d'éléments qui accréditent une exploitation ancienne.

Aussi, à la lumière de cette première lecture technologique qui précède la fouille, trois types de structures ont tour à tour été testées (Fig. 7.6): d'une part, sous le bloc entièrement déchaussé au sommet de l'affleurement (sondage 1); d'autre part, au pied d'un front de taille (sondage 2); enfin, à la base d'un bloc dont l'extraction est inachevée (sondage 3).

Les sondages de 2005

En 2005, trois sondages d'une dizaine de mètres carrés ont précédés la fouille de 2006 qui s'est étendue sur près de 300 m². Les résultats présentés dans cet article sont ceux de 2005. Le sondage 1 a été implanté au pied de l'affleurement nord, sous une paroi verticale dont le sommet est marqué par une petite cuvette d'érosion, et à l'aplomb du bloc de 3,55 mètres de long qui est déconnecté du substrat (Figs. 7.4 et 7.6). Cet emplacement a été choisi afin de recueillir les traces

relatives à l'extraction de ce grand bloc. La fouille a fait apparaître une succession de trois lits homogènes constitués d'éclats ou de moellons en granite emballés dans une terre grise très gravillonneuse et parfois cendreuse, où les tessons de céramique non tournée sont rares.

Hormis quelques gros moellons, les éclats de granite observés dans les différents lits ont une longueur assez homogène comprise entre 5 et 15 cm (Fig. 7.7). Ces pierres ont souvent des arrêtes vives et des traces de rubéfaction associées à un état de surface pulvérulent, des éléments qui plaident pour un chauffage de la roche. Parmi ces éclats thermiques apparaissent des blocs en granite de forme ogivale (Fig. 7.8). Ces blocs se différencient des éclats par une pétrographie différente, des bords arrondis et une absence d'oxydation. Si les éclats thermiques sont de même nature que l'affleurement avec de gros grains de feldspath, en revanche, les blocs de forme ogivale sont peu fissurés et présentent surtout un grain beaucoup plus fin. L'état de surface extrêmement "mâché" à l'extrémité de l'un d'entre eux plaide pour une fonction de percuteur (Fig. 7.8). On découvre dans ces lits des petits éclats de quartz et de grès très anguleux qui sont interprétés comme des fragments de percuteur. Les lits d'éclats de granite situés dans les niveaux inférieurs du sondage montrent des silex brûlés un peu plus nombreux, du charbon de bois en quantité, ainsi que des poches

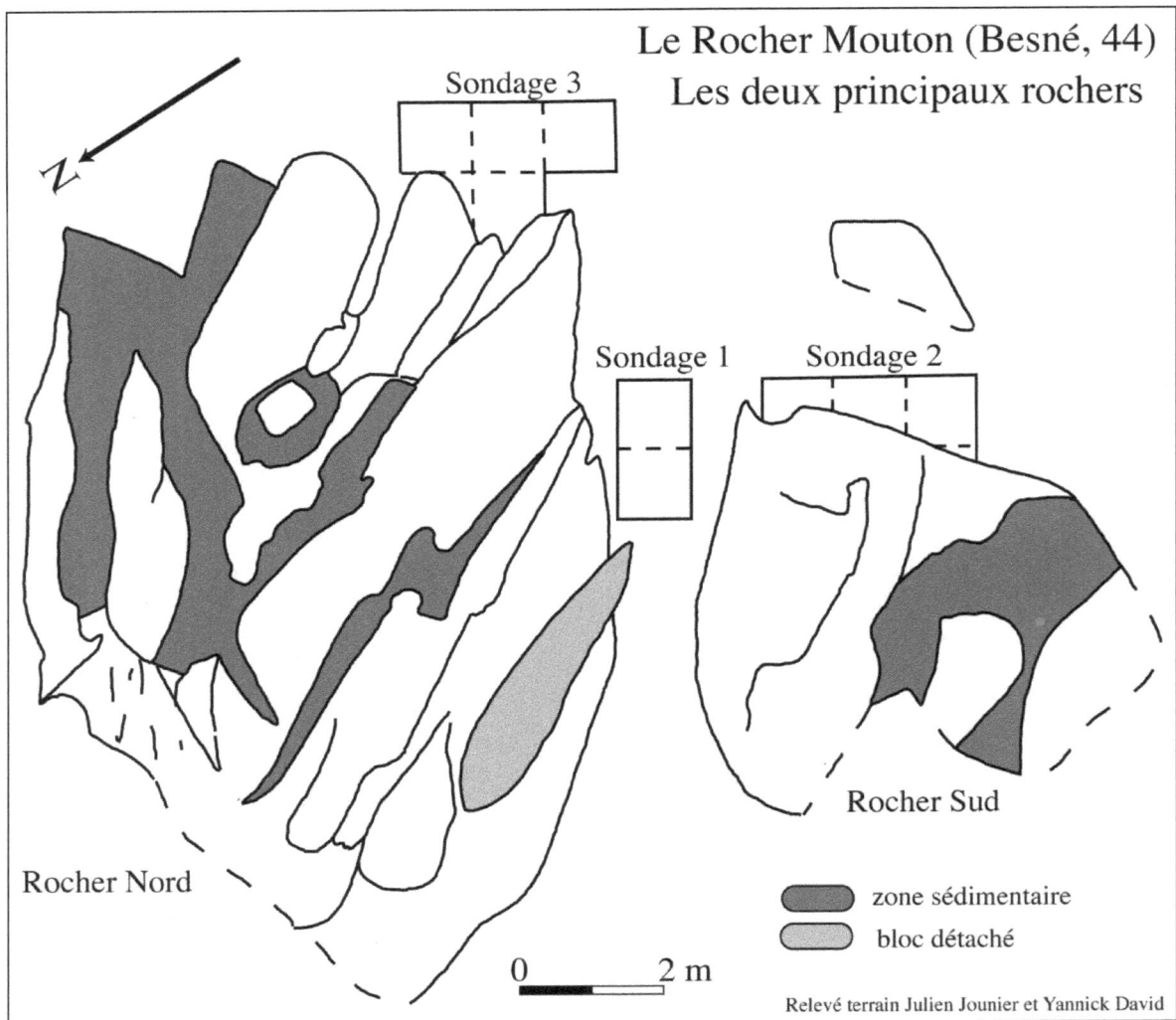

Fig. 7.6. Les affleurements nord et sud du Rocher Mouton

cendreuses. Le substrat apparaît après l'enlèvement du dernier lit de moellons et montre un granite très diaclasé portant des traces d'enlèvements.

En résumé, les trois lits de moellons les plus homogènes du sondage 1 sont apparus dans un horizon gris foncé gravillonneux et cendreux. Les lits sont espacés les uns des autres par un intervalle sédimentaire d'environ 4 cm d'épaisseur. Ces accumulations d'éclats thermiques observées sous forme de lits au pied du rocher associées à des poches cendreuse et du charbon de bois sont interprétées comme des rebuts de carrière, où les percuteurs en granite ont été abandonnés après utilisation. L'emploi de masses en granite pour détacher des éclats de même nature peut surprendre, mais la présence d'éclats thermiques montre que, contrairement aux percuteurs, l'affleurement a préalablement été fragilisé par l'action du feu.

Le sondage 2 s'est installé au pied d'un front de taille choisi en raison de la présence de réseaux polygonaux d'érosion. Ces réseaux polygonaux sont un premier indice

d'ancienneté; de plus, la paroi verticale du front de taille n'a aucune trace d'outils métalliques et montre souvent un état de surface pulvérulent et oxydé.

Dans les niveaux inférieurs du sondage 2 est apparu un horizon brun foncé à gravillons semblable au niveau "gris foncé à gravillons" du sondage 1. Cette couche emballe un lit de moellons dont les mensurations vont de cinq à quinze centimètres de long. Ces moellons sont rubéfiés et associés à des charbons de bois. La fouille a montré au pied de la paroi l'association spatiale entre une diaclase verticale élargie, des éclats, et un percuteur en granite (Fig. 7.9). Comme dans le sondage 1, on retrouve la dichotomie entre des percuteurs en granite à grain fin et des éclats thermiques à gros grains de feldspath. Enfin, le dernier niveau de décapage montre toujours le même horizon brun foncé associé à des gravillons, avec une plus grande quantité d'éclats thermique de taille centimétrique à pluri-centimétrique associés à des charbons de bois. Le socle est atteint rapidement, et on y observe quelques éclats de silex brûlés et une zone de terre très cendreuse collée au front de taille. Cette zone légèrement indurée est

Fig. 7.7. Éclat thermique (photo: A. Oge)

Fig. 7.8. Percuteur en granite (photo: A. Oge)

ponctuée de grains de feldspath, d'éclat de granits brûlés et de charbons de bois.

La stratigraphie du sondage 2 se résume en un horizon humique marron foncé suivi d'une terre brun foncé sans gravillons. On observe ensuite un horizon de terre brun foncé à gravillons avant le substrat granitique. Comme dans le sondage 1, on rencontre dans l'horizon brun à gravillons des éclats thermiques en grande quantité associés à des percuteurs en granite. Tous ces éléments plaident pour la présence de rebuts de carrière provenant du chauffage de la roche pour l'extraction.

Le sondage 3 a été implanté au pied d'un bloc entouré de deux gorges élargies dont l'extraction n'a pas été menée à son terme. Les premiers niveaux de décapage ont fait apparaître un niveau de terre brun foncé à gravillons avec de la céramique non tournée, des éclats de silex, de grès, de quartz, des percuteurs en granite et des éclats thermiques. Tous ces éléments disparaissent dans les niveaux inférieurs où le substrat montre des enlèvements associés à des traces d'outils métalliques. La présence de l'horizon brun foncé à gravillons dans les niveaux supérieurs de la stratigraphie, alors qu'il est caractéristique des niveaux inférieurs dans les sondages 1 et 2, conduit à penser que la

couche préhistorique n'est pas en place dans le sondage 3. La présence de traces d'outils métalliques dans les niveaux inférieurs va dans le sens d'un remaniement des couches préhistoriques lors d'une extraction historique.

La grande majorité des percuteurs découverts au Rocher Mouton dans les sondages 1 et 2 est en granite. Ces percuteurs sont souvent constitués d'un granite à grain fin non oxydé, alors que les éclats de taille sont constitués d'un granite à gros grains de feldspath comme sur l'affleurement. Des masses en quartz et en grès éolisés (Fig. 7.10) ont bien été utilisées, mais elles ont été découvertes en quantité beaucoup moins importante que les percuteurs en granite. L'utilisation de masses en grès ou en quartz a pourtant été significative, comme le montre les nombreux éclats anguleux de petite taille (moins d'un cm^2) découverts lors du tamisage systématique. Ces éclats en grès et en quartz sont présents en grand nombre dans les sondages 1 et 2 dans l'horizon brun ou gris foncé à gravillons. On peut évaluer à quelques centaines de grammes les éclats recueillis dans le sondage 1 (étude en cours). Ainsi, malgré le nombre restreint de percuteurs en grès et en quartz découvert lors de ces sondages, le tamisage fin et systématique démontre que leur utilisation a cependant été importante.

Fig. 7.9. Percuteur en granite (photo: A. Oge)

Fig. 7.10. Percuteur en grès (photo: A. Oge)

La quantité de céramique non tournée découverte en 2005 est faible: à peine une trentaine de tessons retrouvés, principalement dans le sondage 1 (étude Gwenäelle Hamon 2005). Malgré un certain degré d'altération, cette céramique montre des caractéristiques technologiques qui permettent de l'attribuer au Néolithique moyen armoricain (Néolithique moyen 2, 4200-3800 avant J.-C.). Les éléments de comparaison sont plus à rechercher sur des sites du littoral sud-armoricain tels que Er Lannic à Larmor Baden (Morbihan) ou L'Ile aux Moutons à Fouesnant que sur l'habitat proche de la Grande Grée à Sandun (Guérande) (Hamon 2003).

Premier bilan de la fouille d'une carrière mégalithique

On connaissait le rôle considérable joué par les diaclases dans les techniques d'extraction (Gaumé 1992). Le sondage au Rocher Mouton démontre que ces fissures naturelles constituent un critère important dans le choix des sources d'approvisionnement puisque ce site est à la fois le témoin le plus dense en traces d'extraction anciennes et le rocher le plus diaclasé de la région du Bassin du Brivet.

Sur la question des techniques de débitage, les sondages 1 et 2 ont livré une grande quantité d'éclats thermiques qui apparaissent sous forme de lits successifs entrecoupés d'horizons sédimentaires. Selon les lits, les éclats thermiques montrent une grande homogénéité en terme de taille et d'état de surface. Ces lits sont souvent associés à des poches de cendre et à des charbons de bois. Le chauffage de la roche pour l'extraction est également observé sur certains fronts de taille sous forme de "bouffées" d'oxydation associés à un état de surface "pulvérulent". Il apparaît que les percuteurs sont souvent de même nature que le substrat, mais qu'ils s'en différencient néanmoins par une granulométrie beaucoup plus fine. Les percuteurs en quartz et en grès sont moins nombreux; cependant, la découverte d'éclats de même nature grâce à un tamisage systématique montre que leur utilisation a été significative. En l'état actuel des recherches, les accumulations d'éclats thermiques, les percuteurs abandonnés, les éclats en grès et quartz emballés dans un horizon brun ou gris foncé gravillonneux observés au pied des affleurements sont interprétés comme des rebuts de carrière. La présence de percuteurs et de nombreux éclats thermiques au pied des parois oxydées sont autant d'éléments qui plaident pour un travail de percussion effectué sur une roche ayant préalablement subie l'action du feu.

Bibliographie

BENETEAU, G. (2000) *Les alignements de menhirs du Sud de la Vendée*. Editions Anthropologica.

BESSAC, J.C. (1998) L'exploitation à ciel ouvert de la pierre de taille. In *Pierres des monuments, Volume 2: Techniques d'extraction, taille et sculpture*. Paris: Géopré, Maison de la Géologie, 1-13.

GAUME, E. (1992) *Mégalithes et taille du granit à Locmariaquer, de la stèle prédolménique à la margelle de puit*. Mémoire de diplôme de l'Ecole des Hautes Etudes en Sciences Sociales de Toulouse.

HAMON, G. (2003) *Les productions céramiques au Néolithique ancien et moyen dans le nord-ouest de la France*. Thèse de Doctorat, Université de Rennes I-Beaulieu.

HAMON, G. (2005) La production céramique du Rocher Mouton (Besné, Loire-Atlantique). In Mens E., dir, *Rapport de sondage, Le Rocher Mouton, autorisation n°2005-44*. Nantes: Service Régional Archéologique des Pays de la Loire.

LARGE, J.M. (2004) *Deux dolmens de l'île d'Hoedic (Morbihan): redécouvertes et surprises*. Revue Archéologique de l'Ouest, 21, 35-54.

L'HELGOUAC'H, J.; VISSET, L.; SELLIER, D.; PERIDY, P.; BERNARD, J. (2001) L'occupation humaine autour de l'estuaire de la Loire du VIe au IIIe millénaire dans son cadre géomorphologique et paléoenvironnemental". In Le Roux, C.-T., dir, *Du monde des chasseurs à celui des métallurgistes*. Revue Archéologique de l'Ouest, supplément n°9, 9-34.

MENS, E. (2002) *L'affleurement partagé, gestion du matériau mégalithique et chronologie de ses représentations gravées dans le Néolithique moyen armoricain*. Thèse de doctorat en Archéologie, Université de Nantes.

MENS, E. (2005) El remontaje mental del afloramiento: el exemplo del megalitismo armoricano (Francia). In Arias Cabal, P.; Garcia-Monco Pinero, C.; Ontanon Peredo, R., dir, *Actas del III Congreso del Neolitico en la Peninsula Iberica*. Santander: Universidad de Cantabria, 663-669.

MENS, E. (2007) Etude technologique des mégalithes de l'Ouest de la France, les monuments néolithiques du Mané-Bras et du Mané-Bihan à Locoal-Mendon (Morbihan). In Evin, J., dir, *Un siècle de construction du discours scientifique en Préhistoire, Volume 3: Aux conceptions d'aujourd'hui*. Paris: Société Préhistorique Française, 353-9.

PELEGRIN, J. (1995) *Technologie lithique: le Châtelperronien de Roc-de-Combe (Lot) et de La Côte (Dordogne)*. Cahiers du Quaternaire 20. Paris: CNRS.

PRIGENT, D. (1978) *Contribution à l'étude de la transgression flandrienne en Basse Loire: apport de l'archéologie*. Nantes: Association des Etudes Préhistoriques et Protohistoriques des Pays de la Loire.

PRIGENT, D.; VISSET, L. (1977) Etude géologique archéologique et palynologique de deux menhirs situés dans les marais de Brière. *Bulletin de la Société des Sciences naturelles de l'Ouest de la France* 75, 144-161.

SELLIER, D. (1991) Analyse morphologique des marques de la météorisation des granites à partir de mégalithes morbihannais. L'exemple de l'alignement de Kerlescan à Carnac. *Revue Archéologique de l'Ouest* 8, 83-97.

SELLIER, D. (1995) Eléments de reconstitution du paysage prémégalithique sur le site des alignements de Kerlescan (Carnac, Morbihan) à partir de critères géomorphologiques. *Revue Archéologique de l'Ouest* 12, 21-41.

SELLIER, D. (1997) Utilisation des mégalithes comme marqueurs de la vitesse de l'érosion des granites en milieu tempéré: enseignements apportés par les alignements de Carnac (Morbihan). *Zeitschift für Geomorphologie* 41, 319-356.

VISSET, L. (1982) Nouvelles recherches palynologiques dans les marais de Brière: île d'Errand en Saint-Malo de Guersac (Loire-Atlantique, France). *Bulletin de l'Association Française pour l'Étude du Quaternaire* 1, 29-38.

VISSET, L. (1990) *8000 ans en Brière*. Nantes: Editions Ouest-France.

EXPLOITATION DE LA PIERRE ET MISE EN ŒUVRE DES MATERIAUX SUR LE SITE NEOLITHIQUE DU SOUC'H EN PLOUHINEC (FINISTERE, FRANCE)

Michel LE GOFFIC

Conservateur en chef du patrimoine, Centre départemental d'archéologie du Finistère,
16, route de Térénez, 29590 LE FAOU, France

Abstract: Close to the two neolithic cairns of the Pointe du Souc'h are two hollows that have been the subject of partial archaeological excavation. Whereas the slabs of the megalithic tombs are large blocks taken from the nearby seashore, the great mass of smaller stones that form the cairns come from these quarries. The quarrying of the rock is directly related to a system of fractures that run perpendicular to the foliation plane of the orthogneiss and so produces parallel-sided prismatic blocks. The working of the bedrock is attested by the discovery of large stone hammers and mauls consisting of cobbles of various types of rock, some of which weigh 50 kg. Radiocarbon dating of layers of fill dates the quarrying to the Middle Neolithic II period.
Key words: quarry, pebble, rock crack, orthogneiss, stone hammer

Résumé: À proximité des deux cairns néolithiques de la Pointe du Souc'h se trouvent deux dépressions qui ont été partiellement fouillées. Si les dalles des dolmens sont de gros galets prélevés sur l'estran voisin, l'énorme masse des moellons qui constituent les cairns provient de ces carrières. L'exploitation de la roche est liée à la présence d'un réseau de diaclases perpendiculaire à la foliation des orthogneiss qui favorise un débit parallélépipédique. La dislocation du substratum est attestée par la présence de gros percuteurs et masses percutantes qui sont des galets de roches diverses dont certaines atteignent 50 kg. La datation ^{14}C des couches de remplissage situent l'exploitation au Néolithique moyen II.
Mots clés: carrière, galet, diaclase, orthogneiss, percuteur

Si, en ce début du XXIe siècle, les recherches sur l'origine des matériaux ayant servi à ériger des menhirs ou à construire des dolmens et des cairns sont lancées en divers pays, l'idée n'est pas nouvelle. Dans les régions où le calcaire a été utilisé les études sont assez nombreuses; par contre elles sont moins fréquentes en ce qui concerne les granitoïdes, sans pour autant être inexistantes. C'est le cas pour le Massif Armoricain où les références sont plutôt rares et les plus anciennes demandent de sérieux amendements. En effet, dès 1866, Geoffroy d'Ault-Dumesnil publiait un mémoire sous le titre *"Phénomènes de dénudation et de désagrégation, recherches sur la provenance des granits qui ont servi à élever les monuments dits celtiques"*, mais ses conclusions sont entièrement à revoir à la lumière des analyses de matériaux réalisées, de la recherche de leur provenance et ceci en raison des progrès de la géologie et de son application à l'archéologie depuis cent cinquante ans. En 1933, Mourant s'intéressait à la nature et la provenance des matériaux constituant le dolmen de La Hougue Bie à Jersey. Enfin, pour ne citer que trois exemples espacés dans le temps, Pierre-Roland Giot, Louis Chauris et Hervé Morzadec montraient que pour construire le cairn de Barnenez en Plouézoc'h (Finistère), la métadolérite du substratum local avait été utilisée mais aussi des granites circumvoisins, la distance séparant les lieux d'extraction et/ou de ramassage étant inférieure à 2 km (Giot *et al.* 1995). Cependant, avant nos investigations sur le site éponyme de la pointe du Souc'h en Plouhinec entre 2000 et 2006, aucune carrière néolithique n'avait été l'objet de fouille ou de sondage si l'on fait abstraction de celle de Plussulien (Côtes–d'Armor) où la métadolérite a bien été exploitée, mais pour la confection de haches polies (Le Roux 1999). Tout récemment, en octobre 2005, une carrière a été mise au jour à proximité d'un cairn du Néolithique moyen à Saint-Nicolas–du-Pélem (Côtes-d'Armor), mais cette découverte due à Jean-Yves Tinévez est encore inédite.

APERÇU DE LA TOPOGRAPHIE ET DE LA GEOLOGIE LOCALE

Le Cap-Sizun est situé à l'extrémité sud-ouest de la Bretagne et présente au nord une côte à hautes falaises constituant le sud de la baie de Douarnenez et au midi une alternance de plages et de zones rocheuses peu élevées. Au centre, une dépression est empruntée par le Goyen et d'autres rivières qui se jettent dans la baie d'Audierne ou dans celle des Trépassés. L'altitude ne dépasse guère les 100 mètres. Ce relief traduit une différence dans la nature géologique du Cap, avec, au nord, un complexe cristallophyllien composé de micaschistes et de granitoïde – la trondhjémite de Douarnenez – tandis qu'au sud se développe le granite de la pointe du Raz-Quimper au sein duquel se trouve le massif intrusif formé par l'orthogneiss de Porz-Poulhan, daté de 345 ± 8 MA. La dépression centrale correspond à un faillage important connu sous le nom de *Zone broyée sud-armoricaine*.

Le substratum qui affleure par endroits à la pointe du Souc'h en Plouhinec, dans la parcelle 372, est formé par l'orthogneiss œillé de Porz-Poulhan qui peut s'observer sur la côte entre Ménez-Gored en Plozévet et la plage de Guendrez en Plouhinec (Fig. 8.1). Il contient de très nombreux phénocristaux feldspathiques (microcline de 1 à 5 cm) aplatis et étirés dans une trame quartzo-feldspathique (quartz, microcline et albite) claire et

Fig. 8.1. Localisation géographique du site de la pointe du Souc'h et extrait
de la carte géologique de la France au 1/50.000, n° 345, feuille de Pont-Croix

schistifiée, parfois riche en biotite souvent chloritisée et contenant également des muscovites, le cortège de minéraux accessoires étant formé par des apatites, zircons, grenats et opaques. Cet orthogneiss est structuré de façon importante selon des plans orientés N 115° E qui plongent de 60° vers le sud tandis que la linéation plonge de 10° vers l'ouest (BRGM 1981). Il contient deux types d'enclaves homéogènes, l'une mélanocrate, riche en biotite, arrondie, l'autre leucocrate à grain fin, aplatie dans le sens de la foliation. La roche vient souvent à l'affleurement sur les versants sud-ouest et sud du Menez-Dregan et les Néolithiques n'avaient que l'embarras du choix pour s'approvisionner en dalles de gneiss, d'extraction facile puisque nombre de dalles sont dégagées par l'érosion et qu'en bas de versant des dalles sont prises dans des coulées de solifluxion datant de la dernière glaciation. Dans la parcelle 114 se voient deux larges dépressions dont l'origine conjecturale a fait l'objet de la fouille de deux transects de 2 mètres de largeur. Il avait été avancé qu'il puisse s'agir d'effondrements du plafond d'une grotte marine, analogue à celle en cours de fouille dans la falaise voisine (grotte de Menez Dregan), mais ces dépressions en sont éloignées de plus d'une centaine de mètres. Le résultat de la fouille nous permet de dire qu'il s'agit bien de carrières, liées à l'exploitation de la roche pour construire les cairns du complexe mégalithique.

LES CAIRNS DE LA POINTE DU SOUC'H ET LEURS CARRIERES

Connues depuis le début du XIXe siècle, les chambres des dolmens de la pointe du Souc'h ont été fouillées par Claude-Alexis Grenot en 1870 et 1871. Il s'agit en fait de deux cairns comportant chacun plusieurs sépultures, dont des dolmens à chambres compartimentées. Leur histoire est complexe et s'étale sur 2000 ans, au moins. Les fouilles ont été reprises en 2000 par le Service départemental d'archéologie du Finistère et les deux dernières campagnes se sont plus particulièrement intéressées aux matériaux de construction et à leur étude.

La quasi totalité des grosses dalles servant d'orthostates pour la construction des parois des chambres et des couloirs des dolmens proviennent de l'estran proche. En effet, les surfaces et les arêtes des blocs sont très usées, polies et sont le résultat d'une érosion marine.

Une prospection effectuée à marée basse lors de la marée d'équinoxe de septembre 2005 a montré que tout l'estran actuel compris entre Guendrez et Pors-Poul'han est parsemé de dalles aux arêtes et aux surfaces usées par l'érosion marine. Il est vraisemblable qu'au Néolithique, alors que le niveau de la mer était quelque 6 mètres plus bas qu'aujourd'hui, il devait en être de même et que les bâtisseurs des dolmens n'avaient que l'embarras du choix. La distance à faire parcourir à des dalles dont le poids se situe pour la plupart entre une et deux tonnes, quelques centaines de mètres tout au plus, ne représente pas grand chose au regard de certaines prouesses de la même époque, qu'il s'agisse de distance ou de poids ou de difficulté de parcours (montées). En cela, les Néolithiques de la pointe du Souc'h étaient plus opportunistes que ceux de Guennoc (Landéda, Finistère) qui exploitèrent les affleurements voisins des cairns, mais en détachant les dalles du bed rock, comme l'a montré Pierre Gouletquer (Gouletquer 2000). Pour ne citer qu'un exemple finistérien, le menhir de Kerloas en Plouarzel dont la masse est évaluée à plus d'une centaine de tonnes a été transporté sur au moins trois kilomètres en gravissant une pente. L'exemple du menhir d'Er Grah à Locmariaquer, avec ses quelque 280 tonnes est encore plus phénoménal! Par contre l'observation des pierres constituant la masse des cairns montre qu'il s'agit de moellons aux arêtes vives, aux surfaces de diaclase lisses, non altérées et présentant encore très souvent la couleur brun-rouille due à un dépôt superficiel d'hydroxydes ferriques, tel qu'on peut le voir sur les pierres fraîchement extraites de carrière.

Quand on parcourt la lande, entre Pors-Poulhan et Guendrez, un regard averti remarque des affleurements naturels d'orthogneiss dont la hauteur peut atteindre 2 mètres hors sol ainsi que des dépressions non loin du trait de côte. Certaines d'entre elles sont dans le prolongement d'anciennes grottes ou de couloirs d'érosion marine réalisés au cours des millénaires tel le vallon de Poullobos. D'autres sont manifestement l'œuvre de l'homme et correspondent vraisemblablement à des emprunts de pierres pour l'édification de monuments aujourd'hui disparus par l'urbanisation récente de cette partie de la commune. En effet, certains monuments mégalithiques signalés par Le Carguet en 1880 n'ont pu être retrouvés lors de nos prospections. Par contre, de nombreux blocs qui pourraient en provenir entrent dans la maçonnerie de murets de clôture ou dans la confection d'ornements de jardin. Par endroits, des dalles gisent sur le sol et ont manifestement été déplacées. Ainsi en est-il d'un bloc qui se trouve non loin du sentier côtier dans ce qui pourrait être une carrière très ancienne, peut-être néolithique. Nous avons également observé les affleurements de roche soumis à l'érosion météorique depuis des millénaires tout comme l'a fait Horst Schülke pour les granites du pays bigouden (Schülke 1971). Les surfaces de diaclase ont perdu leur aspect lisse et leur planéité, certains feldspaths demeurant plus en relief, la foliation se mettant en évidence et les arêtes sont très érodées ce qui a engendré des formes plus ou moins arrondies (Fig. 8.2).

Un lever topographique (Fig. 8.3) montre clairement que les deux dépressions contiguës qui se trouvent en limite de la rupture de pente, à l'ouest des deux cairns néolithiques, sont oblongues et leurs grands axes ont des directions différentes de celles de la foliation des orthogneiss (N 128 gr) et des diaclases et failles suborthogonales verticales (N 20-40 gr) puisque leurs directions sont N 110 gr et N 150 gr. Les grottes marines et couloirs d'érosion que l'on observe dans la falaise

Fig. 8.2. Affleurement d'orthogneiss dans la lande de la pointe du Souc'h montrant la foliation et le pendage de la roche. On notera l'état d'usure météorique des arêtes et de la surface de diaclase. La mire mesure 0,50 mètre

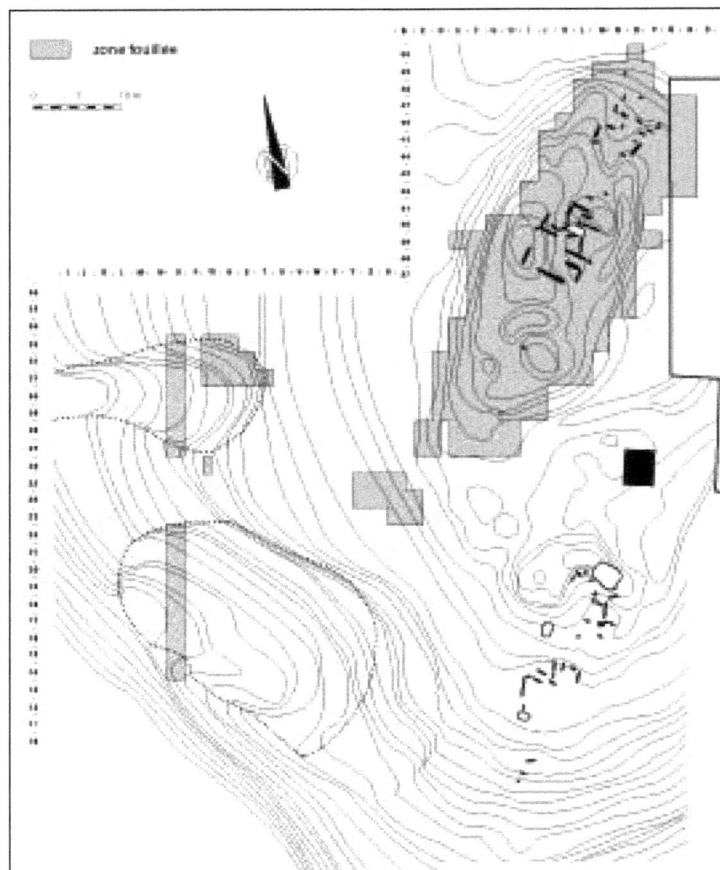

Fig. 8.3. Lever topographique de la pointe du Souc'h mettant en évidence les deux dépressions correspondant aux carrières néolithiques. En grisé, les surfaces fouillées

Fig. 8.4. Fouille du transect O₁ 27-33 mettant en évidence le système de diaclases parallèles et perpendiculaire à la foliation de l'orthogneiss induisant le débit parallélépipédique de la roche

depuis Porz-Poul'han jusqu'à la plage de Guendrez montrent toutes la même orientation générale qui est celle du réseau de diaclases, fissures et failles qui affecte la roche locale et qui l'ont fragilisée quand elles n'ont pas abouti à une mylonitisation. L'hypothèse avancée se trouve ainsi fortement contrariée. Il en résulte que la seconde hypothèse que nous avons émise, c'est-à-dire la présence de carrières liées à la construction des cairns, s'est trouvée confortée par le même fait. Pour la vérifier, deux transects de 14 mètres et de 18 mètres de longueur pour 2 mètres de largeur ont été fouillés.

ETAT DE LA SURFACE DE LA ROCHE

Le dégagement à la truelle et à la brosse de la surface des carrés fouillés a montré un état des surfaces de la roche très frais et des arêtes vives, sans aucune érosion, à l'inverse des affleurements d'orthogneiss observés hors sol. Ceci est révélateur d'une surface rocheuse qui n'a pas été soumise aux agents atmosphériques pendant une longue période. Dans le carré O₁ 33, l'action de la pédogénèse s'est faite sentir et compte tenu que les deux facteurs les plus actifs dans ce type de sol sont l'acidité et le lessivage, dans les sols peu épais (épaisseur inférieure à 0,50 mètre au-dessus de la roche mère), dont le drainage est possible verticalement ou obliquement, comme c'est le cas présentement, la roche apparaît grise et bien lessivée au décapage, sans revêtement ni dépôt argileux. Dans le fond de la dépression, dès que la roche se trouve à plus de 1,10 mètre de profondeur, elle est recouverte d'une fine couche d'argile illuviale qui ne dépasse qu'exception-nellement 5 mm. Le fond de la carrière est tapissé d'un feutrage de radicelles, soit directement sur la roche lorsque la profondeur ne dépasse pas 0,50 mètre, soit sur le revêtement argileux aux niveaux inférieurs. Par ailleurs, de nombreuses diaclases et fissures sont perpendiculaires au plan de foliation des orthogneiss et subverticales (Fig. 8.4). Certaines fissures sont devenues des petites failles qui ont légèrement joué lors de

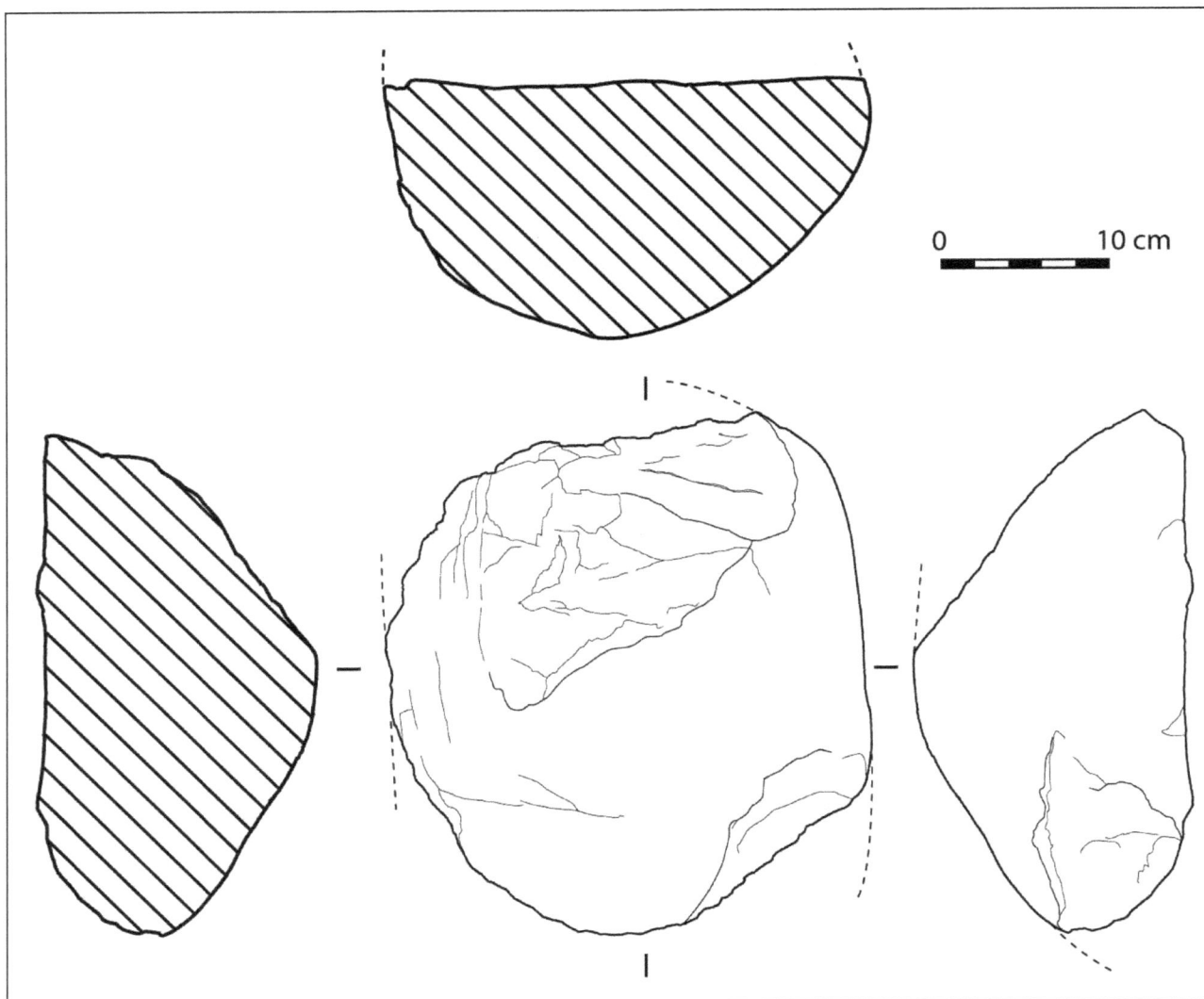

Fig. 8.5. Masse percutante mise au jour dans le transect O$_1$ 14-22, pesant 17,3 kg

secousses postérieures à l'orogenèse hercynienne et sont le témoin d'une néotectonique. L'écart des plans de fissuration est variable et, dans la partie mise au jour, le débit naturel de la roche permet d'obtenir des moellons dont la plus grande longueur n'excède guère 0,50 mètre; les faces de diaclase sont lisses et teintées de brun par les hydroxydes ferriques. Ces pierres sont extrêmement faciles à dégager en raison de la fracturation perpendiculaire à la foliation, mais qui ne permet guère de fournir des blocs de grandes dimensions pouvant servir de piliers ou de dalles de couverture.

Si l'extraction des pierres a pu se faire à l'aide de pics de bois ou d'andouiller de cerf, l'usage de percuteurs est attesté, d'une part par la présence de gros galets montrant des traces de percussion et/ou des enlèvements d'éclats dont certains ont pu être replacés sur le percuteur brisé et abandonné (Fig. 8.5). L'une de ces masses percutantes, dont il manque environ les deux-tiers, est réalisée à partir d'un galet de quartzite et pèse 17,3 kg; c'est donc une masse de plus de 50 kg qui était employée.

Ceci laisse penser que cette masse percutante n'était pas manipulée, mais pouvait se trouver placée à l'extrémité d'une sorte de balancier actionné par plusieurs individus, ce qui augmentait ainsi la force d'application. D'autre part, des traces d'impact ont été relevées sur la roche en place et les nombreux éclats abandonnés et participant au comblement de la carrière sont aussi les témoins de l'usage de la percussion directe. L'étonnement de la roche était superflu dans le cas présent.

LA COUPE DU TRANSECT O$_1$ 27-33

La carrière est longue de 14 mètres et sa profondeur maximum, par rapport au niveau du sol actuel est de 1,50 mètre. La roche arrive pratiquement à l'affleurement à l'extrémité nord de la coupe et apparaît à 0,40 mètre de profondeur à l'extrémité sud. Cette coupe montre que la dépression affecte la forme d'une cuvette allongée dont la profondeur approchée est de l'ordre de 2,50 mètres et la longueur d'une vingtaine de mètres. Le volume de roche

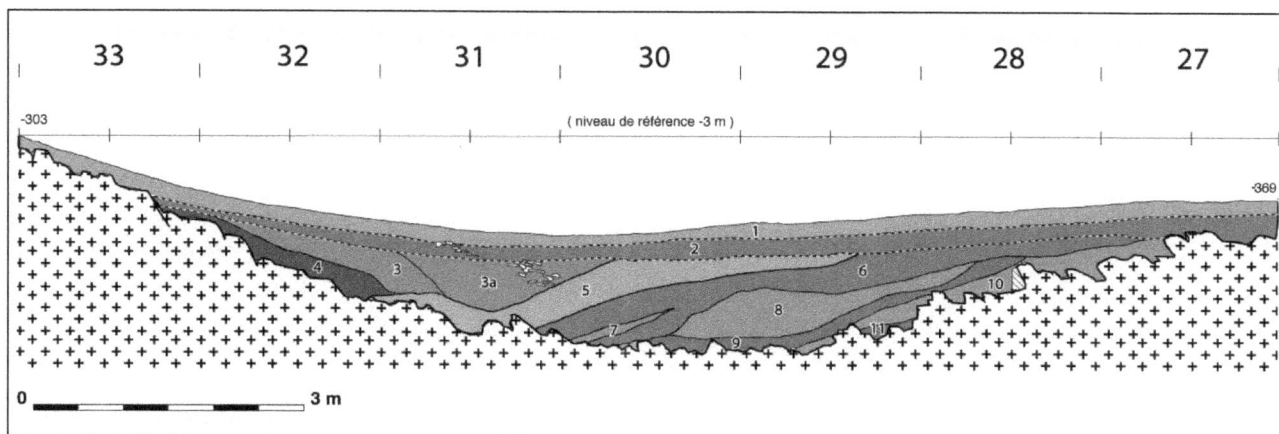

Fig. 8.6. Coupe stratigraphique du transect O₁ 27-33

exploitée est donc proche de 350 m³, soit près de 930 tonnes. Cependant, la totalité de cette masse de pierres n'a pas été exportée car il faut déduire la pierraille laissée sur place et qui contribue au remplissage partiel de la carrière.

La stratigraphie observée est celle du remplissage d'une dépression artificielle partiellement comblée de remblais (Fig. 8.6). La progression du remblaiement s'est faite du sud vers le nord pour la plus grande partie du comblement (US 11 à US 5), le pendage général des couches étant de 27 gr N, avec un maximum de 40 gr N; celui–ci est souligné par l'inclinaison des très nombreux éclats et fragments de roches de module généralement inférieur à 0,10 mètre et qui sont le résultat d'un choix des matériaux. Les couches montrent souvent un infléchissement vers le sommet de la coupe. En fin de remplissage, c'est l'inverse qui s'est produit avec des remblais venant du nord (US 4, 3 et 3a). Le remplissage est constitué d'horizons de sol remaniés et l'on remarque une alternance de couches brunes provenant d'horizons A et de couches de couleur jaunâtre ou rougeâtre venant d'horizons B. À partir d'une cinquantaine de centimètres de profondeur le sédiment est compacté et l'usage du piochon a été nécessaire pour la progression de la fouille. Les unités stratigraphiques sont numérotées de 1 à 11, de la plus récente à la plus ancienne. Notons que c'est sur un remblai que la pédogénèse s'est exercée en fonction des facteurs climatique, biotique et même édaphique du substrat, depuis le Subboréal.

US 1: horizon A₁ du sol actuel, sec, épais de 5 à 15 cm, de couleur gris, comportant des cailloux d'orthogneiss; transition régulière distincte.

US 2: horizon A₂ du sol actuel, frais, épais de 10 à 25 cm, brun foncé, riche en matière organique et comportant également de nombreux cailloux (20 %); transition régulière et graduelle.

US 3a: cette couche représente la dernière phase de remplissage et comprend de nombreuses pierres (30 %) et celles de plus grand module de toutes celles rencontrées lors de la fouille; elles sont emballées dans un sédiment brun rougeâtre.

US 3: sédiment brun-rougeâtre comprenant peu de cailloux de petit module (5 %). A la base, présence de charbons de bois et de nombreux artéfacts: galets, percuteurs, silex.

US 4: sédiment de couleur brun très foncé emballant de nombreux cailloux (50 %) de module compris entre 10 et 20 cm. C'est dans cette US qu'ont été mis au jour les deux seuls blocs utilisables en maçonnerie et de dimensions atteignant 40 à 50 cm. Elles reposent directement sur le bed rock.

US 5: sédiment brun clair rougeâtre riche en cailloux (40 %). La limite avec l'US 3a est assez difficile à percevoir.

US 6: sédiment brun foncé, pauvre en cailloux (5 %).

US 7: sédiment rougeâtre pulvérulent, riche en cailloux et interstratifié dans l'US 6.

US 8: sédiment rougeâtre, pulvérulent, emballant de nombreux cailloux (40%).

US 9: sédiment brun emballant quelques cailloux (10 %).

US 10: sédiment rougeâtre, pulvérulent, emballant de nombreux cailloux (40 %).

US 11: sédiment brun piégé entre deux ressauts d'orthogneiss.

Il est possible de distinguer au moins deux grandes phases de comblement de cette carrière. La première comprend les US 5 à 11 avec un remplissage progressant vers le nord et se termine avec l'US 5; la seconde avec les US 4 et 3 dont le remplissage progresse vers le sud, l'ultime comblement étant réalisé par l'apport de l'US 3a. Ceci laisse penser que le comblement s'est fait pendant l'exploitation de la carrière, les remblais provenant du recul du front de taille, après prélèvement des matériaux intéressants pour la construction des cairns. Les US de teinte foncée proviennent d'horizons A tandis que les US

Fig. 8.7. Carrière nord en fin de fouille avec, en haut du cliché, le front de taille et le réseau oblique de diaclases

ocre jaune ou rougeâtre sont issues d'horizons B plus ou moins spodiques chargés en hydroxydes ferriques.

FOUILLE DE LA CARRIERE NORD, Q_1R_1 31-33, S_1 31-32, T_1 31

De façon à comprendre et vérifier le mode d'extraction de la roche ayant servi à la construction des cairns, la carrière nord a été fouillée sur une surface de 36 m². Dès le décapage des horizons de surface, il est apparu qu'en S_1 31-32 et en T_1 31 le substratum n'avait pas été exploité et que le ranker reposait directement sur l'orthogneiss. Cette fouille a généré deux coupes orthogonales, l'une parallèle à celle levée dans le transect O_1 37-33 et deux mètres plus à l'est, l'autre à la limite des carrés R_1Q_1 30-31.

LE FOND DE CARRIERE ET LE FRONT DE TAILLE

Le décapage des 24 m² du fond de carrière en Q_1R_1 31-33 montre une surface générale assez régulière en cuvette, avec un pendage général vers le sud-ouest, hérissée de petits ressauts de roche aux arêtes vives. Ce qui est remarquable, c'est le système de diaclases et de fissures parallèles qui est perpendiculaire au plan de foliation de l'orthogneiss et subvertical, ce qui induit un débit parallélépipédique en moellons dont les dimensions sont directement en relation avec l'espacement des diaclases et fissures, ce qui avait déjà été noté dans le transect O_1 27-33. L'exploitation de la carrière s'est arrêtée dès qu'un

second réseau de diaclases et fissures à pendage oblique vient recouper le précédent et engendrant ainsi, non plus des moellons parallélépipédiques, mais des pierres de plus petits modules et surtout aux faces se recoupant par des dièdres aux angles aléatoires (Fig. 8.7).

Si, au nord, le fond de la carrière remonte graduellement vers la surface, vers l'est il vient buter contre un front de taille vertical d'une hauteur d'un mètre en moyenne.

Un relevé des directions de diaclases et fissures ainsi que leur pendage a été réalisé sur toute la surface décapée et elles ont été reportées sur un plan (Fig. 8.8). Les mesures en grades et centièmes figurent sur le tableau suivant (Tab. 8.1).

Ces mesures ont servi à réaliser une projection stéréographique des plans de diaclase à l'aide d'une abaque de Wülff (Fig. 8.9). Sur le diagramme, apparaissent nettement quatre familles de faisceaux. La première est orientée N-S (0 à 13 gr) avec un pendage subvertical (91 à 100); c'est celle qui est la plus proche de l'orthogonalité avec le plan de foliation de l'orthogneiss et que l'on retrouve majoritairement dans les carrés Q_1R_1 31-33. La seconde est orientée NE-SO (29 à 55 gr) avec un pendage N-O (68 à 95); il apparaît essentiellement à l'est et au nord de la surface fouillée. La troisième est orientée NNO-SSE (153 à 194 gr) pour un pendage ENE variable de 56 à 95; ce réseau affecte la partie N-E de la zone fouillée. Enfin, la quatrième est faiblement représentée avec une orientation NNO-SSE (160 à 184 gr) mais à pendage accusé O (90 à 100). Une diaclase se

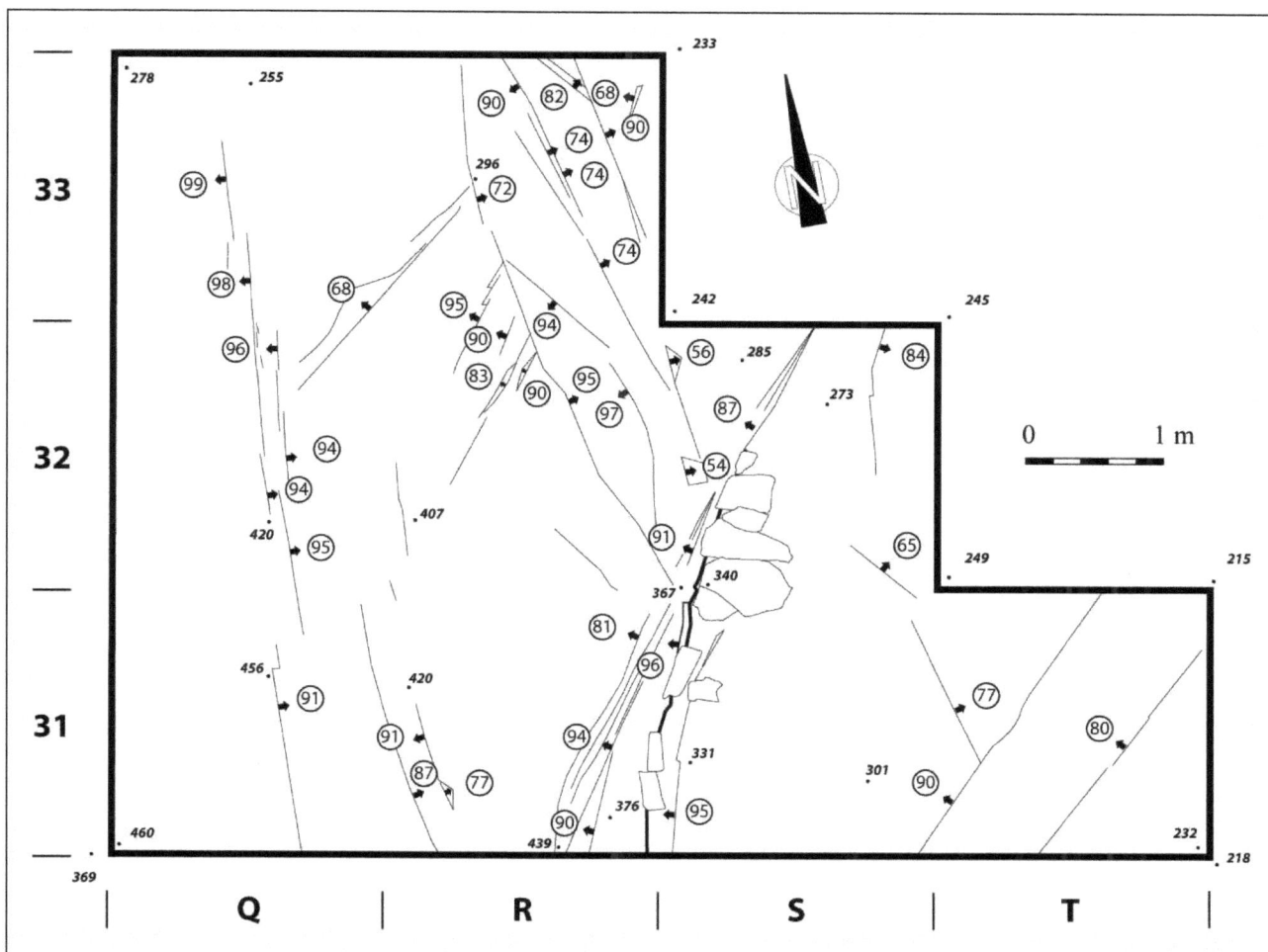

Fig. 8.8. Plan de la partie fouillée de la carrière nord avec les principales diaclases

Tab. 8.1

Azimut	Pendage	Azimut	Pendage	Azimut	Pendage
0.5	91 e	40	95 o	29	89 o
190	87 e	33	90 o	186	95 e
188	91 o	40	83 o	184	97 o
186	77 e	39	90 o	6	97 o
0	95 e	194	72 e	192	56 e
4.5	94 e	6	100	46	87 o
2	94 e	175	90 o	30	84 e
4	100	182	74 e	153	65 e
6	98 o	180	74 e	182	77 e
4	99 o	153	82 e	48	90 o
55	68 o	187	90 e	52	80 o
13	95 o	24	90 o	35	94 o
34	81 o	160	100		

démarque des autres avec une orientation NNE-SSO (30 gr) et un pendage de 84 vers l'ESE. Rappelons que la foliation de l'orthogneiss est orientée à N 128-130 gr pour un pendage de 65 à 75 (croissant en grisé); il est donc manifeste que, compte tenu des différentes orientations des diaclases et de la variabilité des pendages, il devenait

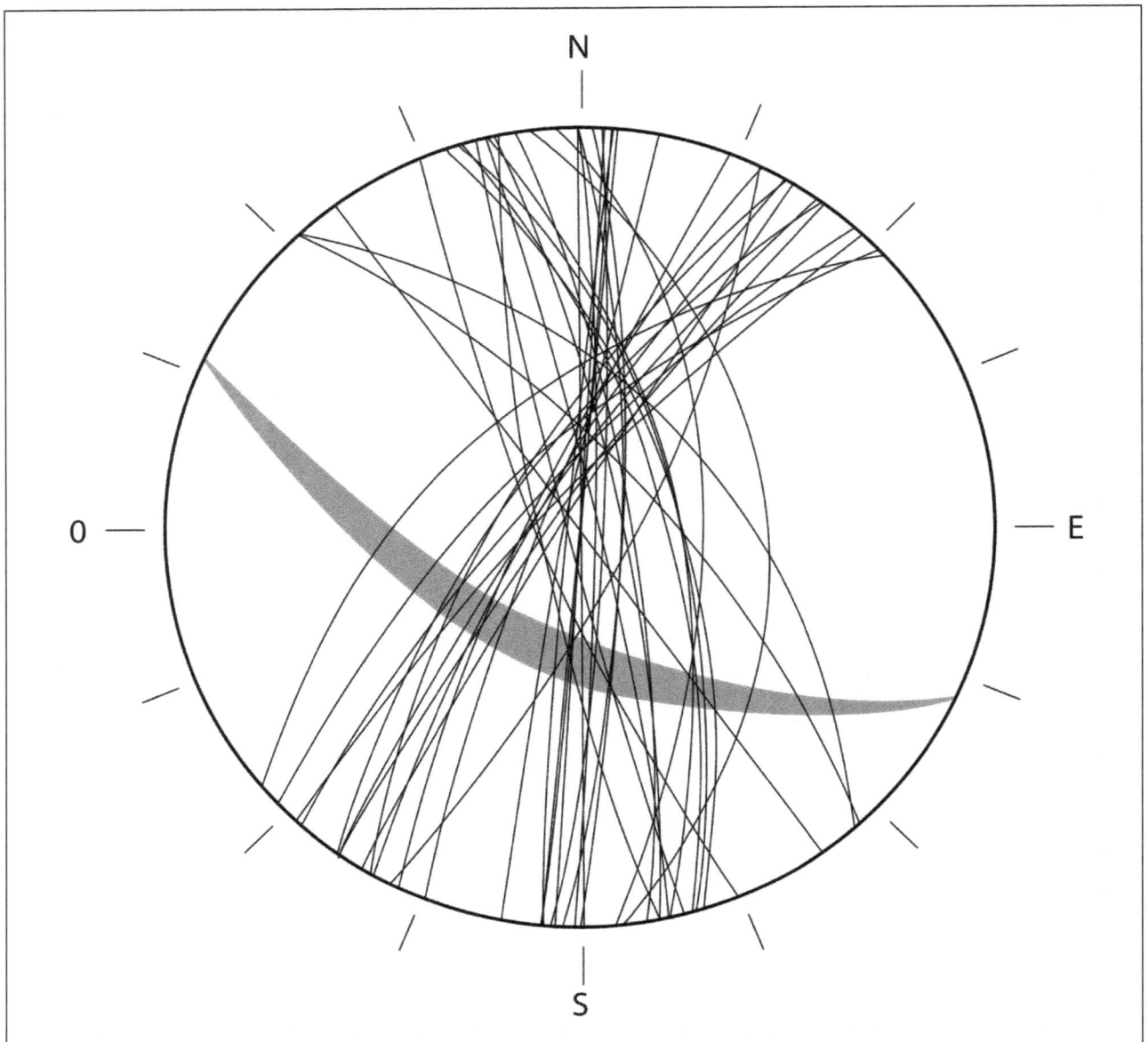

Fig. 8.9. Diagramme de Wülff montrant les familles de diaclases;
le croissant grisé correspond à la foliation des orthogneiss

impossible d'obtenir de moellons parallélépipédiques dans la partie inexploitée de la carrière et c'est bien la raison qui a conduit les néolithiques à en abandonner l'exploitation dans ce secteur.

LES COUPES Q₁ 31-33 ET T₁Q₁ 31

Dans la coupe Q₁ 31-33, on retrouve la stratigraphie de la coupe O₁ 31-33, avec quelques compléments mineurs que nous ne développerons pas ici. Les couches montrent une pente générale vers le nord avec un pendage de l'ordre de 20°.

La coupe T₁Q₁ 31 est perpendiculaire à la précédente et fournit des informations supplémentaires, principalement

dans la zone proche du front de taille où, au-dessus d'une couche de base formée d'une terre de couleur brun-rougeâtre riche en cailloux, se trouve un amas de pierraille plaquée contre le front de taille. Le volume de cet amas correspond à celui de l'abattage de la partie sommitale du front de taille qui se présente en marches d'escalier. Il est recouvert par une couche pauvre en cailloux, de couleur noire, et correspond à un horizon A déplacé. Le mobilier mis au jour dans cette fouille est uniquement lithique (182 objets). Il comprend une quinzaine de galets utilisés dont des enclumes, et sur les 167 silex seulement cinq pièces montrent des retouches; il s'agit d'un grattoir denticulé sur éclat, trois éclats utilisés et un microburin, le reste du mobilier étant du matériel brut. Le tout est attribuable à une occupation mésolithique antérieure à l'exploitation de la roche.

Fig. 8.10. Fond du transect O$_1$ 14-22 montrant une dalle dégagée mais encore en place et une masse percutante réalisée à partir d'un gros galet de quartzite

FOUILLE DU TRANSECT O$_1$ 14-22

Au cours de la campagne 2006, un transect de 18 mètres de longueur a été ouvert dans la seconde dépression, dans le prolongement du premier transect, pour vérifier qu'il s'agit bien d'une seconde carrière. La fouille de la partie médiane du transect a été un peu décevante en ce sens que l'exploitation de la roche a été faible voire nulle en O$_1$ 18, ce qui n'a rien d'étonnant compte tenu que la roche affleure à cet endroit du transect. Le fond de la dépression, à l'inverse de celle précédemment fouillée, est très irrégulier, l'orthogneiss arrivant à l'affleurement au milieu de la coupe ainsi qu'à l'extrémité sud. Le remplissage de la dépression consécutif à la progression de la carrière est moins évident que dans le premier cas. La stratification se fait en couches sensiblement parallèles au fond de la dépression. Plusieurs raisons sont à avancer pour expliquer ces constatations. Tout d'abord, dans cette partie du site le réseau de diaclases est bien moins régulier que dans la partie nord et le débit de la roche ne se fait pas en moellons réguliers. Ensuite la partie sud du transect est affectée par deux failles parallèles dont les plans sont distants de 2 mètres, d'orientation N 165 gr et de pendage NNE 56. Les miroirs de ces failles sont crénelés, en

marches d'escalier, traduisant des rejeux post-tectoniques qui déterminent un débit particulier, orthogonal à ce plan de faille, débit qui a été largement exploité par les Néolithiques. En effet, c'est dans cette partie du transect comprise entre les deux failles que la plus grande profondeur de la carrière a été atteinte. Seule une petite zone mylonitisée, proche de la faille mise au jour en O$_1$ 17, a été négligée car à cet endroit la roche est broyée. C'est aussi dans cette partie du transect, à près de 2 mètres de profondeur, qu'à été mis au jour la plus grosse masse percutante réalisée à partir d'un gros galet de quartzite (voir supra). Comme dans le cas de la coupe du transect O$_1$ 27-33, il n'a pas été trouvé de bloc ou de moellon dans le volume des terres extraites des 36 m^2 fouillés, mais seulement de la pierraille, formée d'éclats de gneiss et de cailloux de formes aléatoires dont le module est décimétrique ou inférieur dans la plupart des cas.

En O$_1$ 16, un bloc a été dégagé par l'exploitation de la carrière (Fig. 8.10). Il est en place et aurait pu former une dalle pour constituer soit un orthostate, soit un élément de couverture. Si l'on considère que la partie encore en place dans le gneiss fracturé se prolonge jusqu'au miroir de faille, ce bloc pouvait avoir dans une de ses dimensions

1,10 mètres, pour une épaisseur de 0,30 à 0,40 mètres. Le plus gros éclat de masse percutante mis au jour se trouvait à la base de ce bloc dégagé. C'est le seul cas de préparation de dalle que nous ayons reconnu dans la fouille des carrières, s'il s'agit bien d'un dégagement intentionnel!

Aucun matériel céramique n'a été découvert dans la fouille de ce transect. Le mobilier lithique compte 651 individus dont la plupart sont des artéfacts mésolithiques. Les seuls microlithes mis au jour sont concentrés dans les trois carrés du nord du transect: en O_1 20, deux trapèzes; en O_1 21, trois triangles scalènes à petite troncature concave; en O_1 22, un scalène et un trapèze. Le reste de l'outillage comprend des lamelles retouchées, des grattoirs frustes, des éclats retouchés ou utilisés auxquels il faut ajouter de nombreux galets utilisés en percuteurs, retouchoirs ou enclumes.

SONDAGE EN Q_1 26 OUEST

Une question restait en suspens: pourquoi existe-t-il deux carrières si proches l'une de l'autre? Une réponse qui vient presqu'aussitôt à l'esprit est qu'elles ne sont pas contemporaines et que les carrières correspondent à des états de construction différents du complexe mégalithique. Une fenêtre a donc été ouverte entre ces deux dépressions, en Q_1 26 ouest. Dès le dégagement des horizons A_o et A qui constituent le ranker à cet endroit de la lande, un filon de quartz laiteux est apparu et l'emplacement du sondage a été judicieusement choisi puisque les deux épontes sont visibles aux angles nord-ouest et sud-est du sondage. L'orientation du filon est nord-est/sud-ouest et sa puissance minimale est de un mètre; autant l'éponte sud est parfaitement nette et rectiligne, celle du nord est plus sinueuse et n'interdit pas l'existence de filonnets secondaires. La mise en place de ce filon de quartz a suivi de peu de temps celle de l'orthogneiss et a subi les contraintes post-tectoniques qu'a connu le Massif Armoricain depuis la fin de l'ère primaire. Il en résulte que le quartz est très fissuré et se débite en tout petits blocs et cailloux qui très vraisemblablement n'intéressaient pas les bâtisseurs des mégalithes du Souc'h. Telle est l'explication que nous proposons pour l'espace non exploité entre les deux carrières néolithiques.

CONCLUSION

Ces observations sur le mode de construction des cairns et dans les carrières de la pointe du Souc'h nous permettent de proposer une gestion très opportuniste des matériaux de construction de la part des bâtisseurs. De plus, nous sommes contraint de constater que ces hommes avaient une excellente connaissance de leur environnement géologique, ce qui leur a permis de se procurer des matériaux de choix pour l'édification des dolmens. En effet, les énormes galets prélevés sur la côte ne nécessitaient pas de dégagement préliminaire et de plus ils avaient subi l'action des assauts de la mer ce qui

garantissait ainsi leur solidité. Les parements des cairns étaient quant à eux montés à l'aide de moellons parallélépipédiques, à arêtes vives et de bonne qualité mécanique, extraits à proximité immédiate des lieux de construction, évitant ainsi tout problème de transport. Une étude qui ne peut être développée dans le cadre de cet article a été entreprise sur les critères typométriques, aspects de surface des pierres en façade, émoussé des arêtes, couleurs et nature lithologique des pierres utilisées. Les résultats croisés montrent que des familles de parements concernant des étapes de construction peuvent être différenciées. Des analyses ^{14}C sur charbons de bois prélevés dans les différentes couches du transect O_1 27-33 ont livré 7985 ± 45 BP (GrA-30245) ce qui donne l'intervalle 7060-6700 cal BC et 7680 ± 45 BP (GrA-30243), soit 6640-6430 cal BC pour les horizons A déplacés lors de la découverture de la roche, ce qui traduit, comme on pouvait s'y attendre au vu du mobilier lithique découvert, une occupation mésolithique du site avant la construction des cairns. En ce qui concerne les couches d'extraction comprenant le rebut de pierraille, deux datations sont quasiment identiques: 4800 ± 40 BP et 4795 ± 40 BP (GrA-30248 et 30249), soit 3660-3380 cal BC, ce qui correspond à une phase de construction des dolmens à chambres compartimentées du Néolithique moyen II et elles corroborent d'autres analyses réalisées sur le cairn septentrional.

Bibliographie

BRGM (1981) *Carte géologique de la France au 1/50.000, feuille de Pont-Croix, IV-19 et notice explicative.* Paris: Bureau de la Recherche Géologique et Minière.

GIOT, P.-R.; CHAURIS, L.; MORZADEC, H. (1995) L'apport de la pétrographie à l'archéologie préhistorique sur l'exemple du cairn de Barnenez en Plouézoc'h (Finistère). *Revue archéologique de l'Ouest* 12, 171-176.

GOULETQUER, P. (2000) Fins de carrières à l'île Guennoc (Landéda, Finistère). In Cassen, S. (dir) *Eléments d'architecture. Exploration d'un tertre funéraire à Lannec er Gadouer (Erdeven, Morbihan). Constructions et reconstructions dans le Néolithique morbihannais. Propositions pour une lecture symbolique.* Chauvigny: Association des Publications Chauvinoises, 555-561.

LE ROUX, C.-T. (1999) *L'outillage de pierre polie en métadolérite du type A. Les ateliers de Plussulien (Côtes-d'Armor): production et diffusion au Néolithique dans la France de l'ouest et au-delà.* Rennes: Travaux du laboratoire Anthropologie, Préhistoire et Quaternaire armoricains.

MOURANT, A.E. (1933) Dolmen de La Hougue Bie: nature and provenance of materials. *Bulletin Annuel de la Société Jersiaise* 12, 217-220.

SCHÜLKE, H. (1971) Le microrelief naturel et anthropique dans le granite du pays Bigouden. *Penn ar Bed* 8, 129-146.

TRANSFORMING STONE: ETHNOARCHAEOLOGICAL PERSPECTIVES ON MEGALITH FORM IN EASTERN INDONESIA

Ron L. ADAMS

Department of Archaeology, Simon Fraser University, Burnaby, British Columbia, Canada V5A 1S6

Abstract: *The relatively recent and, in some cases, ongoing practice of erecting megalithic monuments in eastern Indonesia provides a unique case through which to examine the traditional practice of megalith-building and its associated social and symbolic contexts. Tthis discussion examines how the social and symbolic contexts of megalithic monuments can influence their form. Large stone slabs are quarried, deliberately shaped, and often embellished with elaborate carvings in the construction of stone tombs in West Sumba, Indonesia. This practice contrasts with the erection of megalithic monuments elsewhere in eastern Indonesia, where the shaping of megaliths can be minimal or entirely absent. Comparative analysis suggests that the shaping of large stones and the carving of designs on their external surfaces can be closely tied to the emblematic significance of the stones and their role in the contexts of competition between individuals and groups.*
Key words: *Ethnoarchaeology, Indonesia, megalithic monuments, social competition*

Résumé: *L'habitude relativement récente et, dans quelques cas, encore actuelle d'ériger des monuments mégalithiques dans l'Indonésie orientale nous offre un cas unique permettant d'étudier la pratique traditionnelle de la construction de mégalithes et le contexte social et symbolique qui y est associé. Cette discussion-ci mettent à l'examen les manières dans lesquelles les contextes sociaux et symboliques des monuments mégalithiques peuvent influencer leur forme. De grandes dalles de pierre sont extraites, façonnées à des formes preconçus, et souvent ornées de gravures très soignées lors de la construction de monuments funéraires en pierre dans l'ouest de Sumba en Indonésie. Cette pratique contraste avec l'érection de monuments mégalithiques ailleurs dans l'Indonésie orientale où le façonnage des mégalithes peut être réduit au minimum ou entièrement absent. Une analyse comparative suggère que le façonnage de pierres de grandes dimensions et la gravure de motifs sur leur faces extérieures peut être grandement liée au sens emblématique des pierres et leur rôle dans le contexte de compétitions sociales entre individus et groupes.*
Mots clés: *Ethno-archéologie, Indonésie, mégalithes, compétition sociale*

The social and symbolic roles of megaliths in prehistoric societies have been a major focus of archaeological enquiry, particularly in the analysis of prehistoric social organization. In the past, much emphasis has been placed on the presence of large monuments and their implications for the sociopolitical order of prehistoric societies (e.g. Fleming 1973; Chapman 1981; Renfrew 1976). Given the variation that exists among megaliths, it is understandable that the issue of megalith form has received less attention owing to its variability which may be very contextualized, although more recent discussions have begun to explore the subject (e.g. Bradley 1998; Hodder 1990; McMann 1994; Parker Pearson 1999). These studies have been primarily concerned with the symbolic, and while the variability associated with form certainly warrants a more contextualized and symbolic approach, when dealing with multiple contexts and different types of monuments broader social issues can more easily be addressed. This kind of approach is particularly pertinent when dealing with varying symbolic realms and different types of monuments. In such cases, while specific forms (e.g. dolmens, menhirs) may be influenced by symbolic templates, it is likely that other factors play a role in the size and degree of refinement associated with stone monuments.

The variability that exists in the megalithic traditions of eastern Indonesia represents a unique case in this regard. These large stone monuments vary considerably, especially in the size and modification of the stones. The continued prevalence of megalith-building in the western part of the island of Sumba in particular represents a rare opportunity for ethnoarchaeological research. This paper explores the present-day processes associated with megalith form in West Sumba and extends a similar perspective to megaliths in the Torajan highlands of Sulawesi, Indonesia to examine the relationship between the size and refinement of megalithic monuments and living contexts of power, competition, and social relations. Data for this study was collected during periods of ethnoarchaeological fieldwork conducted by Ron Adams in West Sumba between 2001 and 2005, in addition to work undertaken by Brian Hayden and Ron Adams in Sulawesi between 1999 and 2001. All of this research formed a part of a larger *Ethnoarchaeology of Southeast Asian Feasting Project* directed by Brian Hayden.

WEST SUMBA AND MEGALITHS

The island of Sumba is located in the eastern part of the Indonesian archipelago, immediately south of the island of Flores. Sumba is relatively poor in highly valued trade commodities and has low agricultural productivity by comparison with the wetter and more fertile areas of western Indonesia (e.g. Java, Sumatra and Bali). It has traditionally been a sparsely populated island lacking the kingdoms and states that emerged in other parts of Indonesia. West Sumba, in particular, remains one of the least "globalized" parts of Indonesia, and here megalithic tomb-building, traditional clan social structures, and large

Fig. 9.1. Megalithic tomb in Anakalang, West Sumba (Photo: Ron Adams)

feasts are surviving vestiges of a pre-colonial period (in this case pre-20th century, as can be seen in the observations of Colfs 1888). They are still important aspects of a society that remains tied to the traditional economy based on rice agriculture and domesticated animals.

Large stone tombs in West Sumba are located in the centre of villages that serve as the primary social and ritual locales of lineages and larger clan groups. The commonest element in this burial tradition is the stone slab that serves as the capstone for a tomb. In the simplest of forms, these slabs can rest alone above a grave dug into the ground. More advanced stone-walled megalithic tombs consist of large stone slabs that form the walls with a capstone as the roof. These tombs may be included in the type of stone monument that is referred to as a dolmen, prehistoric versions of which have been found throughout most of western Europe, northern and central Africa, the Arabian peninsula, Madagascar, India, northern China, Korea and Japan, and Colombia (Joussaume 1988, 16-24).

The basic dolmen is the primary tomb form in the traditional domain of Kodi, on the western coast of Sumba. Megalithic dolmens in Kodi usually consist of five stone slabs (four walls and one cap or roof) that form a structure or room in which the remains of the deceased are interred. Such tombs are typically about one and a half

metres tall, two metres long, and about one and a half metres wide, although considerably larger versions may reach weights of approximately 30 tonnes or more. The more elaborate stone tombs in Kodi are adorned with carvings on the outer wall faces and on the sides of the capstone. In addition, some tombs have free-standing stones (about one and a half metres in height) at one or both ends of the tomb.

Yet more sophisticated tombs are found in other parts of West Sumba. These consist of a stone box tomb underneath a large stone table-like structure with four or six legs (about one metre tall and fifty centimetres wide) and a stone slab top that may be two to three metres long and one to two metres wide. It is common to find a large free-standing rectangular stone a couple of metres in front of these tombs (Fig. 9.1). This standing stone, along with the legs and the edge of the slab forming the table top, are usually elaborately carved. Such large and decorative stone tombs represent the grandest megalithic monuments in the Anakalang area of West Sumba and were traditionally reserved for the wealthiest members of the nobility, while several smaller and less elaborate styles of tomb (e.g. simple dolmens or simple stone slabs) were built by those of lower socio-economic standing.

The construction of tombs can be an extensive process beginning with the quarrying (Fig. 9.2), which entails the

Fig. 9.2. Removing a large stone slab from a quarry in Kodi, West Sumba (Photo: Ron Adams)

labour of a specialized stone quarrier and his assistants. The quarrying group, consisting of 10 to 40 individuals, is typically contracted with livestock (one water buffalo, one horse, and one pig) and finely woven cloth as payment. In addition, the owner of the stone must provide the quarry crew with food (meat and rice) while the stones are being quarried.

Quarrying the stones for a tomb can take from about one week to a month or more. During this time, each stone is dug from the ground and chiselled smooth to the proper dimensions required for building the tomb. In some cases, particularly in Kodi, special capstones are quarried from Tarimbang, a village in the eastern part of Sumba where an especially desirable type of fine-grained sandstone is found. Individuals in West Sumba must contract a crew from Tarimbang to have these stones quarried and transported to West Sumba, which traditionally entailed a journey by boat and cost of 20-30 head of livestock (a combination of water buffaloes and horses).

The traditional method for transporting megalithic stones in Sumba is to haul them atop wooden sledges (*tena watu*) (Fig. 9.3). A sledge with its attached stone is pulled using vine ropes, requiring between 100 and 1000 people for the large capstones or standing *kado watu* stones, while lesser numbers are needed to move the stones for the tomb walls (50-100 people). It can take from one day

to nearly one month to transport the largest stones in this manner from a quarry to the tomb owner's village, depending upon the size of the stone and the distance to be travelled. Each day the stone is moved, several pigs, and at times water buffaloes, are slaughtered to feed the stone hauliers and the spectators who are invited to view the proceedings. The labour for this endeavour typically comes from the tomb owner's clan as well as from other allied clans.

Once all of the stones have arrived at the tomb owner's village, they are assembled and sometimes carved with elaborate designs on their external surfaces by a specialized stone carver (Fig. 9.4). Such carvings may add an additional month to the process and require further payments of livestock to the stone carver and his assistants. The addition of these carvings to the exterior of tombs further enhances their display value. Lavish tombs may be adorned with several types of carvings. In Kodi, the motifs on a tomb exterior can be direct symbols of wealth, such as the traditional prestige items of gongs, gold earrings and water buffalo horns, which are said to illustrate the wealth of the individual who owns the tomb. In some cases the owner's name or even his genealogy is carved on the tomb (Hoskins 1986, 39). In other parts of Sumba, symbols of descent, succession and ritual precedence (all of which are linked to the overall prestige of the individual) also adorn tombs and the standing

Fig. 9.3. Hauling a large stone for a tomb in Anakalang, West Sumba
(Photo: Ron Adams)

Fig. 9.4. Carving designs on the exterior of a tombstone in Anakalang, West Sumba
(Photo: Ron Adams)

Fig. 9.5. Slaughtering a pig for a tomb-building feast in Anakalang, West Sumba (Photo: Piter Rehi)

stones that can be found in front of the tombs (Hoskins 1988). Recent innovations can also be found. These include water buffalo horns modelled in cement protruding out from the tomb, cement figures dancing atop the tombs, painted decoration, a miniature stone version of a large lineage house built atop a tomb, and the tiling of the tomb exterior. Unlike the entoptic imagery carved inside the Neolithic passage graves of southern Brittany (Lewis-Williams & Dowson 1993, 60), these images of wealth and power are meant to be seen by a wide audience.

There are several factors to consider when assessing the forms of these monuments. From a symbolic standpoint, megalithic tombs in Sumba are considered "houses of the dead" (*alli mate* in the Kodi language of the western coast of Sumba). As houses of the dead, particularly those of prominent individuals, a certain level of refinement should be expected. Large lineage houses in West Sumba are impressive in their design and in the carvings that adorn their internal posts. The tombs that stand in front of these houses are similarly well-constructed and have a visual impact when viewed from the feasting plaza that they surround. Although these tombs are by no means replicas of houses of the living, the designation of tombs as houses is not unlike the "metaphorical model" put forth by Parker Pearson (1999, 37, 38) in which he suggests that early stone tombs in southern Madagascar were

modelled on houses. This link between tombs and houses could also lend support to the notion that the forms of European Neolithic long mounds were derived from long houses (Hodder 1990; Bradley 1998).

Beyond their association as the houses of the dead, the visual impact of tombs in West Sumba is said to relate to the perceived wealth of the tomb owners (who can build their own tombs while still living) and the prominence of their clan. The visual significance of tombs is not only tied to the grandeur of large tombs, but to the heavy expenditures of time, labour, and resources required for their construction. The feasts necessary to feed those individuals who took part in quarrying, hauling, and carving the stones may require well over 100 pigs (Fig. 9.5) and more than 10 water buffaloes for slaughter (and even greater expenses can be incurred when transporting stones from different parts of the island, such as Tarimbang, as noted above). Moreover, details concerning the numbers of livestock slaughtered for building large tombs are often passed down through generations within a family. Indeed, building a tomb can give increased access to the inner circle of clan power through the special title that is conferred upon those who have built tombs, held large feasts, and sponsored the construction of large lineage houses. It is these individuals who become the revered ancestors of their clans.

While these tombs are a reflection of the renown of individuals, they are also to a lesser extent a reflection of the clan group structures in which these individuals claim membership. Tombs in Kodi are considered not only the property of the individuals who acted as the primary sponsors of their construction, but also are considered to be owned by the person's affiliated clan. The dynamics of tomb-building in West Sumba are closely tied to clan group structures, since tomb construction requires support in terms of food and labour from a network of individuals from within one's clan.

In many parts of West Sumba, clans remain highly relevant sociopolitical units, not only for their support in building tombs. While individual households hold use rights to particular segments of land, all land within a clan is traditionally considered to be collectively owned by the clan as a whole. Cooperative labour within clans is used for housebuilding and agricultural work. In addition, clans provide important political support in disputes with other clans and in accessing modern administrative posts. Participation in clan undertakings (including megalith-building) is considered essential for maintaining a voice in clan affairs and accessing these clan support networks. In this way, the tombs reflect the environment of competition between clans that was traditionally manifested in inter-clan warfare and is currently reflected in periodic land disputes between neighbouring clans. Megalithic quarry sites are also clan-owned property and expenditures of livestock are required for use of a quarry belonging to another clan. The clan connection of megaliths is further signified by the fact that megalithic tomb slabs quarried for another clan are considered to be like a bride who is marrying into that clan (Hoskins 1986, 33). This link between tomb-building and corporate group structures echoes Chapman's (1981) conception of megaliths in the Neolithic of western Europe acting as signifiers of land-use rights and corporate group power, although in West Sumba megaliths are not considered to be markers for land rights *per se*.

The presence of an older and simpler tomb-building tradition in Sumbanese villages suggests that tombs may in the past have had a stronger link to group structures and have not always been associated with wealth and power. In Kodi villages, older tombs exist that are simple stone slabs over a burial in the ground. The stones are generally unmodified or have undergone very little modification. Villagers claim that these tombs were part of a burial tradition predating the dolmens and that people retrieved the unmodified stones for these tombs from the coastline adjacent to the areas where limestone is currently quarried. According to informants, the simplicity of these tombs results from the lack of the technology that would be required for quarrying (e.g. iron tools). While this explanation is impossible to verify with the current archaeological data from Sumba, it is worth noting that the names of those interred in these simple tombs in Kodi are not remembered. In addition, there is very little space between each slab, which together

reportedly overlie a large number of interments. This may signify an earlier burial tradition that was less individualistic and focused more on the solidarity of group-oriented clan structures, in contrast with the current dolmen tradition in which the names of the individuals interred in even the oldest tombs are remembered for several generations. Similar single-slab tombs in the traditional domain of Anakalang in the eastern part of West Sumba are considered to be the tombs of commoners and the degree of elaboration is directly related to the higher social standing of the tomb-builder as a member of the nobility. What this all points to is a general correlation between tomb elaboration, including morphology, and the degree to which tomb-building is tied to traditional power structures. That is, the largest and most refined tombs are a reflection of the wealth and sociopolitical prominence of the living tomb-builder.

MEGALITHS OF TANA TORAJA, SULAWESI

Extending the geographical scope, large stone menhirs were traditionally constructed in the Torajan highlands of the Indonesian island of Sulawesi, about 600 km north of Sumba. Besides the obvious difference between tombs and menhirs, the level of refinement associated with menhirs in Tana Toraja also differs from that of the megaliths in West Sumba. Torajan menhirs, although often shaped, typically do not exhibit the same level of elaboration as the tombs on Sumba and designs were traditionally not carved on their exteriors. These stones were sometimes hewn minimally or not at all depending upon the shape of the stone in its natural state, the desired shape typically being a long, slender monolith that tapers at one end. In the Sangalla area of south-central Tana Toraja, however, stones with outstanding natural shapes, such as a bird, were chosen for use as monuments without undergoing any modification (Nooy-Palm 1979, 267). In all cases, megaliths in Tana Toraja were traditionally procured from hills and streams, rather than being quarried from specific quarry sites (Chrystal 1974, 120). The contracting of specialized quarry crews and the expenses associated with feeding such crews is thus absent from the Torajan megalithic tradition.

In addition, in contrast to the situation on Sumba where individuals build their own tombs while still living, megaliths in Tana Toraja are erected by the deceased's surviving relatives. These descendants act as the primary sponsors of the funeral for which a megalith was traditionally erected. The expense that may be associated with funerals is so great that they are often held many years after the deceased has passed away. The largest Torajan funerals can reach excessive proportions in terms of their duration (several days) and the number of water buffaloes slaughtered (which can be well over one hundred). Torajan funerals may also be associated with spectacles such as buffalo fighting, cock fighting and the erection of temporary structures to house guests. In one recent case, an entire hotel was built on the occasion of a

Fig. 9.6. Menhir at a Torajan *Rante'* (feasting plaza) (Photo: Brian Hayden)

funeral to house guests who had travelled some distance to attend. These funerals represent significant feasts at which important relationships are established and enhanced through debt creation (i.e. reciprocal obligations generated by contributing livestock for feasts), success is promoted (especially by the sponsors of the funeral), and power is accessed (particularly in the administration of traditional labour organizations) (Adams 2001).

Despite the differences between the megalithic traditions of West Sumba and Tana Toraja, the overall social contexts of the monuments in these areas show significant similarities. The size of the megaliths in Tana Toraja is considered to reflect the wealth and prominence of the deceased and the deceased's kin group (Crystal 1974). As in West Sumba, hundreds of hauliers from within and outside of one's kin group may be required to pull the large Torajan menhirs using vine ropes. The numbers needed to transport moderately-sized menhirs (40 individuals reportedly moved a menhir some 4 metres long and one-third of a metre in width: Ames 1998, 111) are smaller, however, than those associated with even modestly-sized tombs in West Sumba, where the capstones may demand well over 100 hauliers. All the menhirs of a kin group (*tongkonan*) are situated together in a feasting plaza (Fig. 9.6), which parallels the location of tombs around clan ceremonial areas in West Sumba. The relationship between a desired form and the socio-economic prominence of the sponsors also appears to be common both to Tana Toraja and West Sumba. While the Torajan megaliths do not receive the same degree of modification as those in West Sumba, the effort is made to obtain a certain shape either through searching the landscape for suitable stones or through carving. The smaller and 'misshapen' are in both cases the least attractive options.

DISCUSSION

This assessment of the different levels of elaboration associated with the megalithic traditions of Tana Toraja and West Sumba reveals a complexity that is not easy to explain. The lesser modification of the stone (i.e. absence of quarried blocks and carved designs) in the Torajan area could perhaps be viewed as part of a symbolic metaphor in which humanity is a part of nature rather than actively dominating it through land-clearance and large-scale agriculture (Tilley 2004, 85-86). The traditional context of Tana Toraja could not, however, be considered any less developed from an agricultural standpoint than that found in West Sumba. By all indications, Tana Toraja was traditionally more complex socially, politically, and economically than West Sumba, with greater social differentiation and more complex social institutions as well as closer connections with neighbouring larger kingdoms (Adams 2001). In this regard, less elaborate stone monuments need not reflect a less complex social scenario.

The religio-symbolic contexts of megalith-building in West Sumba and Tana Toraja are also similar, both regions having experienced major conversions from indigenous animistic traditions to Christianity. In West Sumba, the construction of large tombs has continued in spite of their association with the indigenous *marapu* religion in which tomb building represents the journey (or future journey) of the deceased's soul. In Tana Toraja various aspects of funerals, such as the construction of wooden *tau tau* effigies representing the deceased, were eliminated from Christian funerals by missionaries (Adams 1993, 62; Volkman 1985, 131). The erection of menhirs, which are sometimes considered to house the deceased's soul, has largely been absent from Christian funerals in the latter half of the twentieth century since they conflict with Christian conceptions of the place of the soul after death (Ames 1998, 111). These prohibitions have not, however, completely prevented the erection of either *tau tau* or menhirs at certain funerals (Ames 1998, 125). Importantly, the large-scale slaughter of domesticated animals that stands at the core of the sociopolitical dynamics that are played out between individuals and groups at funerals in Tana Toraja has remained a feature of Christian funerals and has even been amplified during the course of the twentieth century. Thus religious factors

alone do not appear to explain the lesser emphasis on stone monuments in Tana Toraja.

In West Sumba, the competition between individuals and groups associated with tomb-building in the prestige economy appears to have led to a greater focus on megalith-building than in Tana Toraja. This is reflected in the special title accorded to renowned tomb-builders and feast-givers in West Sumba. In Tana Toraja, the large funeral that includes the erection of a megalith represents a similar process of competition between individuals and kin groups seeking access to labour and sociopolitical influence. These funerals have increased in their elaboration in terms of the numbers of guests who participate and the numbers of livestock that are slaughtered, despite the general absence of megalith-building (Adams 2001). In Tana Toraja, the lavish expenditure associated with funerals, leaving aside the erection of stone monuments, may always have made megalith-building less significant in terms of the investment put into the creation of elaborate forms. That could explain why megalith-building has become a very rare occurrence in Tana Toraja, while in West Sumba, megalith-building remains a prominent practice today.

In both cases, there are strong indications that the refinement of megaliths and the level of investment associated with megalith-building is related to the importance of these monuments as signifiers of power and renown. The question then becomes, what non-megalithic material signifiers have survived in Tana Toraja that might indicate a social context of competition similar to that in West Sumba? Perhaps the best indication of this lies in the architecture of large kin-group houses (*tongkonan*). These structures represent the ancestral houses of Torajan kin-groups and have elaborate painted designs carved on their exterior and high-peaked roofs (Waterson 2000, 182-183; Adams 2001, 26-27). The lineage houses (*uma*) of West Sumba represent similar group structures, although the architecture of these buildings is noticeably less lavish with a general lack of carvings on the exterior walls and an absence of embellished architectural designs. Even without the megaliths as material evidence of large funerals, the horns of water buffaloes slaughtered at funerals that are displayed on the exterior of Torajan *tongkonan* are clear vestiges of a kin group's feasting (Nooy-Palm 1979, 232, 234). The traditional erection, consecration, and rebuilding of these structures are displays of prominence similar to tomb-building in West Sumba. Thus while megaliths can clearly be symbols of power and renown, specific attention must be directed to their form and to the larger picture of material displays within societies when seeking to determine their social significance.

CONCLUSIONS

This ethnoarchaeological perspective on the subject of megalith form in eastern Indonesia gives several prominent insights into factors that can influence the degree to which stone is modified in the creation of monuments. Megalith-building in West Sumba occurs in a context of competition between individuals and clans for resources and access to sociopolitical influence, which can be gained through constructing large stone tombs. Such competition appears to be linked to the size and refinement of megaliths, since the size and grandeur of the megaliths tends to be associated with the wealth of the tomb builder and the perceived renown of their clan. Extending a similar perspective to the megaliths in the Torajan highlands of Sulawesi reveals some important similarities. In both cases, stone monuments have display properties related to the prominence of individuals and groups and are built in contexts in which power is achieved and negotiated at large feasts. The different degrees of elaboration associated with megalith-building on Sumba and Sulawesi appear to result from the different ways in which megalith-building and other material displays are related to power and competition.

Acknowledgements

Brian Hayden's direction and support has been essential to the completion of the research presented in this paper (the data for which was collected during the course of my MA and PhD theses at Simon Fraser University). Others who have played a role in the work include Piter Rehi, Thomas Tedawonda, Agusthinus Galugu, Pak Agustinus Sabarua, Pak Rehi Pyati, Pak Octavianus Ndari, Suzanne Villeneuve, Umbu Siwa Djurumana, Janet Hoskins, and many informants in West Sumba and Tana Toraja, Indonesia. The fieldwork in Indonesia was funded by the Social Sciences and Humanities Research Council of Canada and was conducted in collaboration with Dr Haris Sukendar (Pusat Penelitian Arkeologi Nasional, Jakarta) and Ayu Kusumawati (Balai Arkeologi Denpasar) for work in West Sumba. Work in Tana Toraja was conducted in collaboration with Dr Stanislaus Sandarupa (Universitas Hasanuddin). Field research in Indonesia was undertaken with permission from the Indonesian Academy of Sciences (LIPI). Thanks also are extended to Michèle Caron for the French translation.

Bibliographie

ADAMS, K., (1993), The discourse of souls in Tana Toraja (Indonesia): indigenous notions and Christian conceptions, *Ethnology* 93, 55-68.

ADAMS, R., (2001), *Ethnoarchaeology of Torajan Feasts*, Unpublished MA Thesis, Burnaby, British Columbia: Department of Archaeology, Simon Fraser University.

AMES, T.T., (1998), *Feasting on Change: The Impacts of Modernization and Development upon the Toraya Traditional Roles, Rituals, and Statuses*, Unpublished PhD Thesis, Burnaby, British Columbia: Sociology and Anthropology Department, Simon Fraser University.

BRADLEY, R., (1998), *The Significance of Monuments: On the Shaping of Human Experience in Neolithic and Bronze Age Europe*, London: Routledge.

CHAPMAN, R., (1981), The emergence of formal disposal areas and the 'problem' of megalithic tombs in prehistoric Europe. In Chapman, R., Kinnes, I., & Randsborg, K. (eds.), *The Archaeology of Death*, Cambridge: Cambridge University Press, 71-81.

COLFS, A., (1888), *Het Journaal van Albert Colfs: eene bijdrage tot de kennis der kleine Soenda-Eilanden*, A.G. Vorderman (ed.). Batavia: Ernst.

CHRYSTAL, E., (1974), Man and the menhir: contemporary megalithic practice of the Sa'dan Toraja of Sulawesi, Indonesia. In Donnan, C.B. & Clewlow, C.W. Jr. (eds.), *Ethnoarchaeology*, Los Angeles: Institute of Archaeology, University of California, Los Angeles, 117-128.

FLEMING, A., (1973), Tombs for the living, *Man* 8, 178-193.

HODDER, I., (1990), *The Domestication of Europe: Structure and Contingency in Neolithic Societies*, Oxford: Blackwell.

HOSKINS, J., (1986), So My Name Shall Live: stone-dragging and grave-building in Kodi, West Sumba, *Bijdragen to de Taal-, Land- en Volkenkunde* 142, 31-51.

HOSKINS, J., (1988), Arts and Cultures of Sumba. In Barbier, J.P. & Newton, D., (eds.), *Islands and Ancestors: Indigenous styles of Southeast Asia*, New York: te Neues Publishing Company, 120-137.

JOUSSAUME, R., (1988), *Dolmens for the Dead: Megalith-building throughout the World*, London: Batsford.

LEWIS-WILLIAMS, J.D. & DOWSON, T.A., (1993), On vision and power in the Neolithic: evidence from the decorated monuments, *Current Anthropology* 34, 55-65.

McMANN, J., (1994), Forms of power: dimensions of an Irish megalithic landscape, *Antiquity* 68, 525-544.

NOOY-PALM, H., (1979), *The Sa'dan-Toraja: A Study of Their Social Life and Religion 1*, The Hague: Martinus Nijhoff.

PARKER PEARSON, M., (1999), *The Archaeology of Death and Burial*, College Station: Texas A&M University Press.

RENFREW, C., (1976), Megaliths, territories and populations. In De Laet, S.J. (ed.), *Acculturation and Continuity in Atlantic Europe*, Brugge: De Tempel, 198-220.

TILLEY, C., (2004), *The Materiality of Stone: Explanations in Landscape Phenomenology*, Oxford: Berg.

VOLKMAN, T.A., (1985), *Feasts of Honor: Ritual and Change in the Toraja Highlands*, Urbana and Chicago: University of Illinois Press.

WATERSON, R., (2000), House, place, and memory in Tana Toraja (Indonesia). In Joyce, R.A. & Gillespie, S.D. (eds.), *Beyond Kinship: Social and Material Reproduction in House Societies*, Philadelphia: University of Pennsylvania Press, 177-188.

www.ingramcontent.com/pod-product-compliance
Lightning Source LLC
Chambersburg PA
CBHW051306270326
41926CB00030B/4736